Letts

Revise GCSE

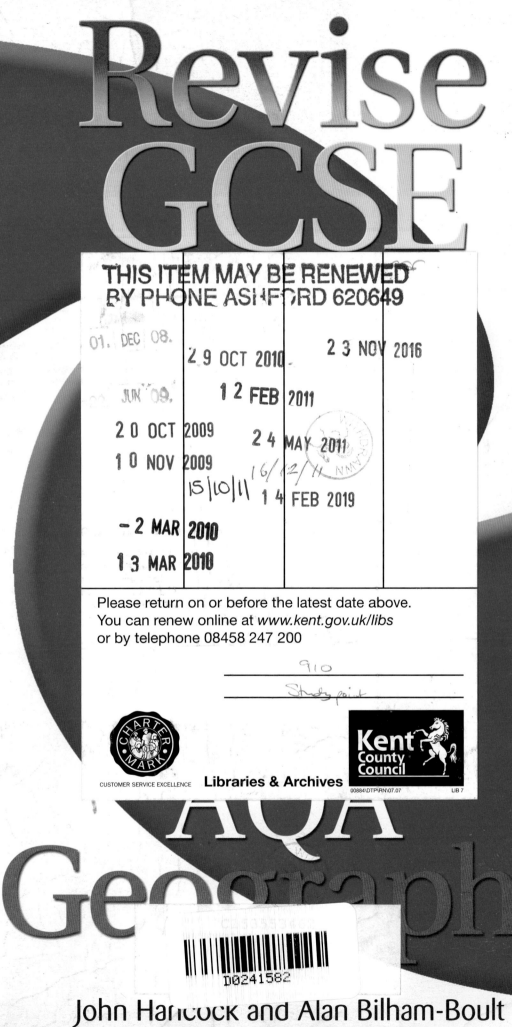

AQA Geography

John Hancock and Alan Bilham-Boult

Contents

Preparing for the examination

Planning your study

The final **three months** before taking your GCSE examination are very important in achieving your best grade. However, the success can be assisted by an organised approach throughout the course.

- After completing a topic in school or college, go through the topic again in your *Revise GCSE Geography Study Guide.* Copy out the main points again on a sheet of paper or use a highlighter pen to emphasise them.
- A couple of days later try to write out these key points from memory. Check differences between what you wrote originally and what you wrote later.
- If you have written your notes on a piece of paper, keep this for revision later.
- Try some questions in the book and check your answers.
- Decide whether you have fully mastered the topic and write down any weaknesses you think you have.

Preparing a revision programme

In the last three months before the final examination go through the list of topics in your Examination Board's specification. Go through and identify which topics you feel you need to concentrate on. It is a temptation at this time to spend valuable revision time on the things you already know and can do. It makes you feel good but does not move you forward.

When you feel you have mastered all the topics, spend time trying past questions. Each time check your answers with the answers given. In the final couple of weeks go back to your summary sheets (or highlighting in the book).

How this book will help you

Revise GCSE AQA Geography Study Guide will help you because:

- It contains the **essential content** for your GCSE course without the extra material that will not be examined.
- It contains **Progress checks** to help you to confirm your understanding.
- It gives **advice from an examiner on how to improve your answers**.
- The summary table will give you a **quick reference** to the requirements for your examination.
- **Marginal comments** and **highlighted key points** will draw to your attention to important things you might otherwise miss.
- It contains important **case studies** to supplement those used in your course and these will help you gain higher marks.

Six ways to improve your grade

1. Planning your time

Make sure you **know how many questions** you have to answer and if there are any rules about which ones are compulsory and which include choice. **Plan how much time** you have for each question. Keep to time – do not answer too quickly or too slowly as both can lead to lost marks.

2. Read the question carefully

Many students fail to answer the actual question set. Perhaps they misread the question or answer a similar question they have seen before. **Read the question once right through and then again more slowly.** Some students underline or highlight key words in the question as they read it through. Questions at GCSE contain a lot of information. You should be concerned if you are not using the information in your answer. Command words are usually the first words in a question. They are very important because they tell you exactly what the examiner wants you to do. When you read the question for the first time underline the command word e.g. **describe**, **explain**, **compare**.

3. Give enough detail

If a part of a question is worth **three marks** you should make at least **three separate points** or develop one fully. Be careful that you do not make the same point three times. Approximately 25% of the marks on your final examination papers are awarded for questions requiring longer answers. Practise writing eight to twenty lines of text to answer a question worth 8 to 20 marks.

4. Quality of Written Communication (QWC)

From 2003 some marks on GCSE papers are given for the quality of your written communication. This includes correct sentence structures, correct sequencing of events and use of geographical words.
Read and correct your answer through slowly before moving onto the next part. Remember, you must finish the paper.

5. Using geographical terms correctly

There is an important geographical vocabulary you should use. Try to **use the correct geographical terms** in your answers and spell them correctly. The use of the right terms in the correct place is important in showing your understanding of the topic. Use the right term when labelling diagrams. Using the right term saves time in the examination. As you revise, make a list of geographical terms you meet and check that you understand the meaning of these words.

6. Case studies are important

Many answers gain higher marks when you **use a case study of a real place to illustrate the points you are making**. Make time in your revision to learn them carefully, include local place names and if possible draw a sketch map.

Useful websites

Rocks and landscapes

education@dartmoor-npa.gov.uk is the website of Dartmoor National Park and has fact sheets on geology and landforms (including 'tor' formation), evolution of landscape, relief, soils, vegetation and farming

www.bbc.co.uk/education/rocks offers an animated interactive timeline about continental drift, fossils and rocks. Archive programmes include Dartmoor, Portland stone (Dorset) and the Lake District

River landscapes and hydrology

www.curriculumvisions.com has an excellent unit on rivers

www.oxfam.org.uk/whatnew/features/banglada.htm is a case study of flooding in Bangladesh

Coastal landscapes

www.georesources.co.uk/leld.htm has information on landforms, erosion and longshore drift, as well as a case study on Recluver (Kent)

www1.npm.ac.uk/lois/Education/case.htm has a case study about the Holderness coastline

www.swgfl.org.uk/jurassic has information about fossil collecting at Lyme Regis and other towns in the area of the Jurassic Coast World Heritage Site

Glacial landscapes

www.antarctica.ac.uk is the website of the British Antarctic Survey

glacier.rice.edu has information on glaciers and ice sheets in Antarctica

Weather and climate

www.met-office.gov.uk is the official website of The Meteorological Office and has education sections as well as weather forecasts

www.worldclimate.com holds 85 000 records of world climate data

www.newscientist.com/channel/earth/climate-change is part of the New Scientist magazine website and contains special reports on climate change

Ecosystems

www.ucmp.berkeley.edu/glossary/gloss5/biome/index.html explains and describes the world's major biomes

www.eduweb.com/amazon.html is the Amazon Interactive website, containing information and activities about the Ecuadorian rainforest

Tectonic activity

www.usgs.gov is the website of the United States Geological Survey and gives information on world earthquakes and volcanoes

volcano.und.nodak.edu/vw.html (Volcano World) contains everything you could want to know about volcanoes

http://www.fs.fed.us/gpnf/volcanocams/msh is a webcam focused on Mount St Helens volcano in Washington State, USA

Population

www.statistics.gov.uk provides up-to-date national statistics for the UK

www.populationconcern.org.uk is a helpful website for the study of population issues

www.populationconcern.com focuses on world news about population

Settlement

www.unhabitat.org is a United Nations website describing their programme for development control

www.bartonhillsettlement.org.uk deals with a specific inner city project in the UK

www.shelter.org.uuk is a charity website concerned with housing for the poor and homeless

www.georesources.co.uk provides case studies on settlement issues

Energy

www.bwea.com gives information on British wind energy development

www.worldenergy.org is concerned with sustainable energy development at a world scale

www.dti.gov.uk/energy is a government website concerned with the production and supply of energy

Agriculture

www.globalgiving.com deals with the issues of agriculture and sustainability

www.ukagriculture.com provides a useful guide to farming

www.usda.gov represents the US Department of Agriculture

Industry

www.nmsi.ac.uk is the website of the National Museum of Science and Industry

www.tandf.co.uk is a website for those interested in industrial issues

www.bbc.co/2/ni/science is a BBC news website about industrial matters

Development, trade and aid

www.dfid.gov.uk is a government website promoting international development

www.edrd.com provides a European view on development

www.odi.org.uk promotes sustainable development

www.actionaid.org.uk is a global antipoverty agency

www.georesources.co.uk provides accounts of issues related to resources and case studies

www.greenpeace.org.uk provides news on resource issues around the world

www.unep.org is the United Nations website for development issues

Tourism

www.tourismconcern.org.uk provides an ethical and sustainable view of tourism

www.visitwales.com gives information on tourism opportunities in Wales

www.tourismtrade.org.uk provides free quarterly updates on the tourist trade

www.culture.gov.uk is the official website for government support of tourism

This book and your GCSE course

Make sure you know which syllabus you are studying.

The section headings used in this book match those in AQA syllabuses A and C. Students following syllabus B will find these section headings support their case studies and provide coverage of the syllabus. In many places the case studies reflect those found in Syllabus B.

	AQA A	AQA B	AQA C
Specification Number	3031	3032	3033
Specification Structure	3 sections all examined	4 regions at different scales, all examined	3 sections, all compulsory
Terminal Papers	Paper 1 1 hr 45 min 40% Skills and physical Ans. 3 Qs from 7 Paper 2 1 hr 30 min tiered, 35% Human and economic 3 Qs from 6 Pop & Settlement Agriculture & Industry Managing resources & Development	Paper 1 1 x 1 hr 15 min 30% UK + OS map All Qs compulsory Paper 2 1 x 2 hr tiered, 45% EU, world and global All Qs compulsory	Paper 2 1 x 1 hr 45 min tiered, 50% Human, physical, economic 3 Qs, all compulsory
Coursework	25%, 2500 word fieldwork investigation	25%, 2500 word fieldwork investigation	25%, 2500 word fieldwork investigation
DME	None	None	Paper 1 1 x 1 hr 30 min all Qs compulsory
People and the Natural Environment			
Rocks and landscapes	9.2, 11.1, 11.2, 11.3	9.1, 14.1, 14.2, 14.3	9.8, 10.1, 10.2, 10.3
Hydrological cycle and rivers (incl. flooding)	9.4, 11.1, 11.2, 11.3	9.1, 9.2, 14.1, 14.2, 14.3	9.6, 9.8, 10.1, 10.2, 10.3
Coasts (incl. management)	9.5, 11.2	11.2, 12.3, 14.2	9.8, 10.2
Ice (incl. management of upland glaciated area)	9.3, 11.2, 11.3	9.2, 9.3, 14.2, 14.3	9.8, 10.2, 10.3
Earthquakes and volcanoes (incl. plate tectonics)	9.1, 11.2, 11.3		9.5, 10.2, 10.3
Weather and climate (incl. micro-climates and tropical storms)	9.6, 11.2, 11.3	9.2, 11.1, 11.2, 14.2, 14.3	9.6, 10.2, 10.3
Climate change (incl. global warming, desertification, acid rain, ozone depletion)	9.6, 11.2, 11.3	12.3, 14.2, 14.3	9.7a, 9.11, 10.2, 10.3
Ecosystems (incl. management and sustainability)	9.7, 11.2, 11.3	11.1, 14.2, 14.3	9.10a, 10.2, 10.3
People and the Human Environment			
Population	10.1	11.2, 11.3	9.1, 9.2
Settlement	10.2	9.6	9.3, 9.2
Urbanisation	10.2	9.6, 12.1	9.3, 9.4
Energy and resources	10.5	9.4	9.10
Sustainability	10.5	11.1, 12.3	9.11
Agriculture	10.3	9.2	9.9
Industry	10.4	9.5	9.9, 9.11
Industry in MEDCs	10.4	10.4	9.9, 9.11
Industrial change	10.4	9.5	9.9, 9.11
Development, trade and aid	10.6	10.1, 10.2, 10.3	9.9
Tourism	10.5	9.3, 10.3	9.12
Geographical skills (incl. OS map)	11.1, 11.2, 11.3	14.1, 14.2, 14.3	10.1, 10.2, 10.3

Visit your awarding body website for full details of your course or download your complete GCSE specifications.
www.aqa.org.uk

Chapter 1

Rocks and landscapes

After studying this section you should understand:

- *how to describe a physical landscape*
- *the nature and formation of igneous, sedimentary and metamorphic rocks*
- *the different types of weathering*
- *the role of rocks and weathering in three different landscapes*
- *the importance of and issues related to the quarrying industry*

Describing the shape of a landscape

A **landscape** consists of a group of **landforms**, which need to be described. Landscapes can be described in the field with reference to a map, and by drawing and labelling a field sketch. A landscape can also be described from a photograph. The following factors should be considered:

- **height** (altitude) in metres, of a number of prominent points or areas
- **relief** – the general shape of the land, the differences in height between the lower and higher areas
- **slope of the land** – steep, gentle, flat
- **shape of the upland areas** – jagged, craggy, smooth
- **shape of the valleys** – flat-floored, U-shaped, V-shaped, gorges
- **nature of the surface of different areas** – rock outcrops, loose rock fragments, soil-covered
- **presence of water** – lakes, sea, rivers (meanders), waterfalls
- **vegetation** – moorland, forest, woods, grass, marshes
- **human impact** – quarries, roads, villages, isolated farms, fields
- **personal attitude and feelings** – attractive, awesome, desolate, tedious, exciting.

Practise reading the contours on a map. Look at the key. What is the contour interval?

If contours are close together, the slope is steep.

A V-shaped valley has Y-shaped contours pointing upstream.

Fig 1.1 Annotated field sketch of the Nant Francon Valley in Snowdonia, North Wales (ref OS 1:50 000 map no. 115)

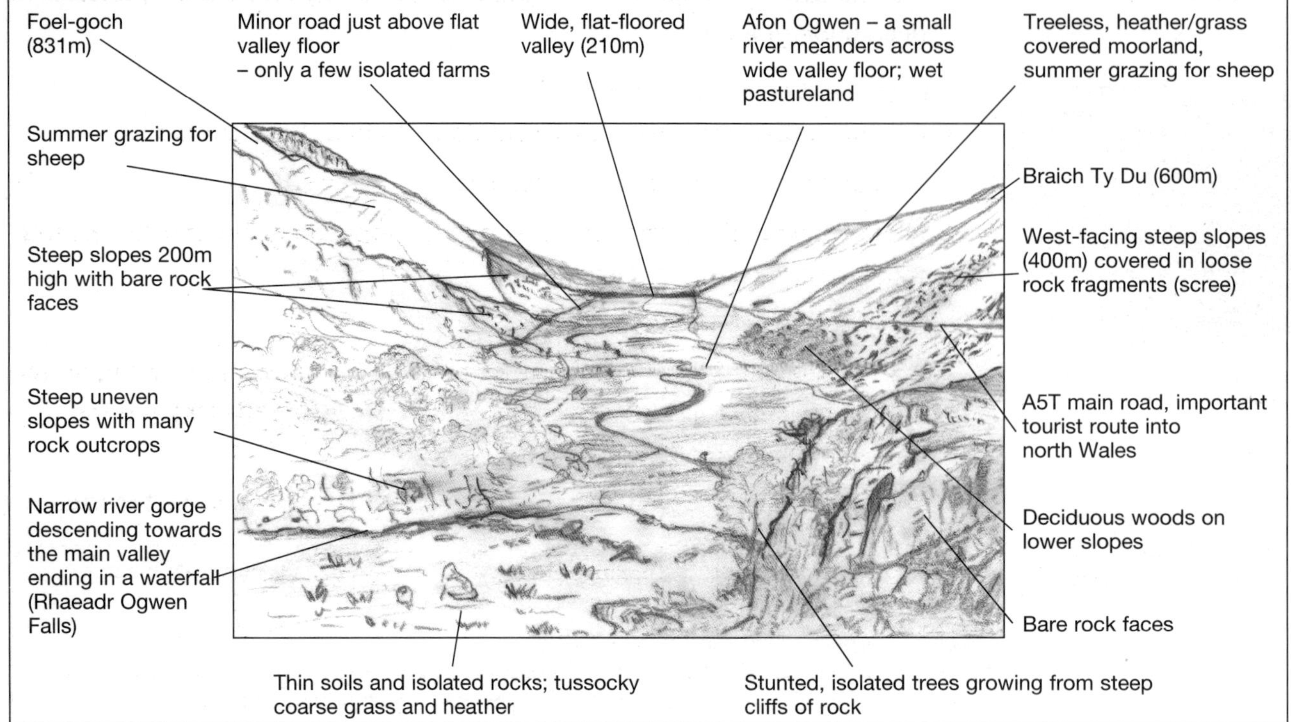

Explaining how a landscape has been formed

Understanding and explaining the formation of a landscape, and the variety of landforms that are found there, must take the following factors into account.

- The types of rock.
- The structure of the rocks.
- How different rocks vary in their resistance to:
  - weathering processes
  - the geomorphic processes of erosion, transportation and deposition, resulting from: the work of rivers; the work of the sea; the work of ice; tectonic forces.
- How in the past, when the climate was colder or warmer, weathering and geomorphic processes were different from those affecting the landscape today.

The formation of rocks

 AQA A AQA C

Rocks can be divided into three types based on their formation:

- igneous
- sedimentary
- metamorphic

Igneous rocks

- **Igneous** rocks form from the molten interior of the Earth. They can be divided into two types.
- **Volcanic** rocks form when molten material, such as lava, is thrown out onto the surface of the Earth and cools quickly.

Fig 1.2 Granite

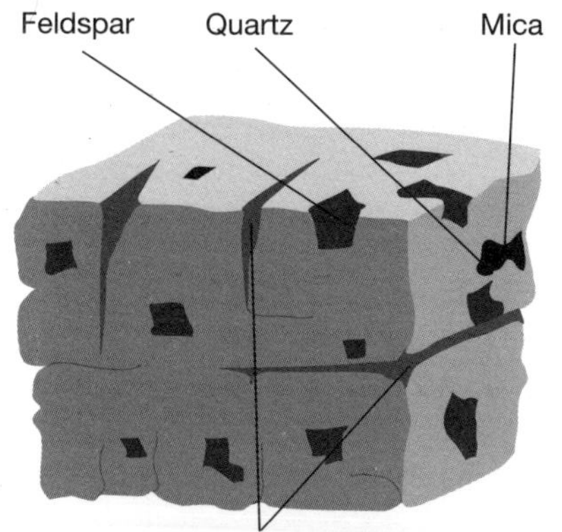
Feldspar
Quartz
Mica

Joints and cracks are the result of pressure release as the rock comes closer to the surface

Understand these terms clearly and use them in your answer.

- **Intrusive** rocks, such as **granite**, form when large masses of molten material are forced up into the crust, and cool slowly under great pressure deep underground.

Some features of granite

- Minerals within the molten rock cool slowly to form large crystals of:
  - quartz
  - mica
  - feldspar.
- Joints, cracks and weaknesses develop as the rock comes closer to the Earth's surface, and pressure is released.
- Granite is **impermeable**.
- Granite is a hard resistant rock that forms upland areas such as Dartmoor, the Cairngorm Mountains in Scotland (a centre for skiing in the UK) and the dramatic cliffs at Land's End.
- Weathered granite produces **kaolin** used in china, paper and toothpaste.

KEY POINT

An impermeable rock allows very little water to pass through it, most rainwater stays on the surface and runs off.

A permeable rock allows water to:

- **penetrate into the rock through the pore spaces – porous rock**
- **pass along joints and bedding planes – pervious rock.**

Sedimentary rocks

Fig 1.3 Sedimentary rock

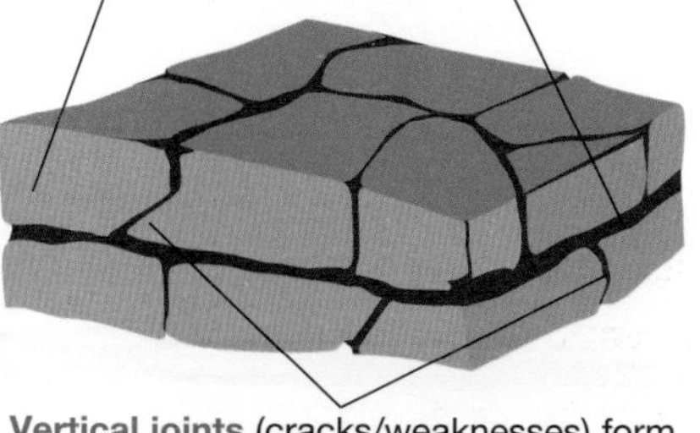
Deposited in layers or beds of rock of varying thickness
Beds are separated by horizontal bedding planes
Vertical joints (cracks/weaknesses) form in some rocks as they dry out and harden

Sedimentary rocks form when rock, which has been eroded by rivers, ice or the sea, is transported away and **deposited**. Limestone, chalk, sandstone and clay are all types of sedimentary rock.

Structure of a sedimentary rock

Some examples of sedimentary rock:

- Carboniferous limestone:
  - a hard grey rock composed of calcium carbonate
  - has prominent bedding planes and many joints
  - very **permeable (pervious)** with little surface drainage
  - forms upland areas such as the **Yorkshire Dales**
  - used as a building stone, in cement and steel making
- Chalk:
  - a white or grey rock with many fine pore spaces
  - has many bedding planes and joints
  - very **permeable (porous)** with little surface drainage
  - used in cement making
  - large amounts of underground water are stored in the pore spaces in the rock called an **aquifer** – important source of water for homes and industry
  - forms hills such as the **North and South Downs** and the **Chilterns**

- Clay:
  - formed of very fine particles of rock
  - impermeable with lots of surface drainage (marshes, rivers and streams)
  - many river valleys are eroded into clay, e.g. Oxford clay vale from Oxford to Peterborough
  - used for brickmaking, pottery

Metamorphic rocks

Metamorphic rocks form when rocks such as sedimentary rocks are altered by pressure or heat in the crust deep below the Earth's surface. Clay rocks can change to become **slate** as found in north Wales. Limestone can change to become **marble** as found on Iona, north-west Scotland.

Weathering – the processes of rock disintegration

KEY POINT **Rocks at or near the surface of the Earth are broken up or decomposed *in situ* by weathering processes.**

Rock that has been weathered or broken into fragments is more easily eroded by rivers, ice or the sea and transported away.

There are three types of weathering:

- **physical** (mechanical)
- **chemical**
- **biological**

Understand clearly the difference between weathering and erosion.

Physical weathering

Freeze-thaw or frost shattering

Fig 1.4 Stages in a freeze-thaw cycle

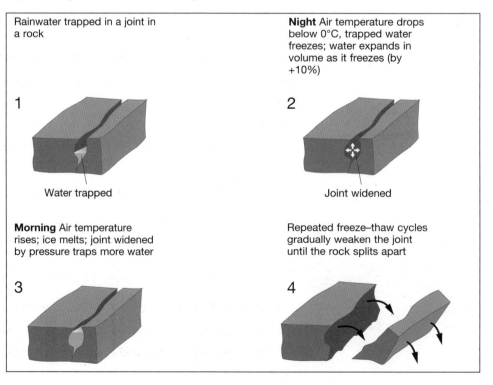

1 Rainwater trapped in a joint in a rock — Water trapped

2 **Night** Air temperature drops below 0°C, trapped water freezes; water expands in volume as it freezes (by +10%) — Joint widened

3 **Morning** Air temperature rises; ice melts; joint widened by pressure traps more water

4 Repeated freeze–thaw cycles gradually weaken the joint until the rock splits apart

Fig 1.5 Cross-section of a scree slope

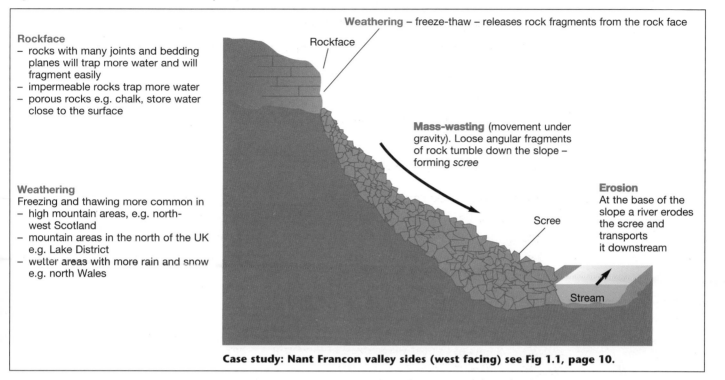

Rockface
- rocks with many joints and bedding planes will trap more water and will fragment easily
- impermeable rocks trap more water
- porous rocks e.g. chalk, store water close to the surface

Weathering
Freezing and thawing more common in
- high mountain areas, e.g. north-west Scotland
- mountain areas in the north of the UK e.g. Lake District
- wetter areas with more rain and snow e.g. north Wales

Weathering – freeze-thaw – releases rock fragments from the rock face

Rockface

Mass-wasting (movement under gravity). Loose angular fragments of rock tumble down the slope – forming *scree*

Scree

Erosion
At the base of the slope a river erodes the scree and transports it downstream

Stream

Case study: Nant Francon valley sides (west facing) see Fig 1.1, page 10.

Fig 1.6 Diagram of limestone section

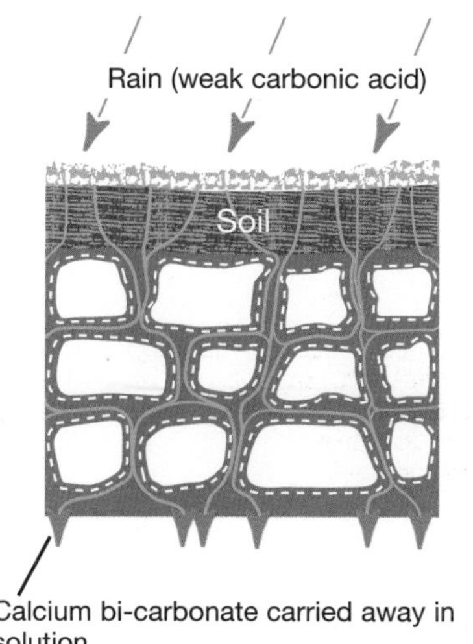

Fig 1.7 Diagram of limestone pavement

grykes, deep widened joints

clints, upstanding blocks

Pressure release

Rocks such as granite cool deep in the Earth's crust, under very high pressures. When exposed at the surface, as on Dartmoor, the pressure has been released and many horizontal and vertical joints and weaknesses have developed in the rock. (See Figs 1.8 and 1.12)

Chemical weathering

Rainwater passing through the atmosphere absorbs carbon dioxide and becomes a weak carbonic acid. The acid reacts with minerals (chemicals) forming the rock, causing decomposition.

Limestone solution (carbonation)

- Limestone is formed of insoluble calcium carbonate.
- Weak acidic water, passing through the joints, reacts with calcium carbonate converting it to calcium bi-carbonate, which is soluble and carried away in solution.
- Chemical reactions were faster during periods in the past when climates were warmer than today.
- Joints and bedding planes are slowly widened.

If the soil is removed, a surface is left of upstanding blocks separated by wide joints – **limestone pavement**.

Granite disintegration (hydrolosis)

- Granite is partly composed of the mineral feldspar.
- Weak acidic water passing along joints and weaknesses reacts with the hydrogen ions in the feldspar, which decomposes into clay (kaolin).

Fig 1.8 Granite disintegration (hydrolysis)

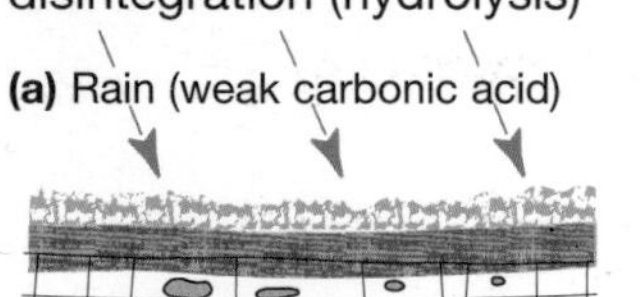

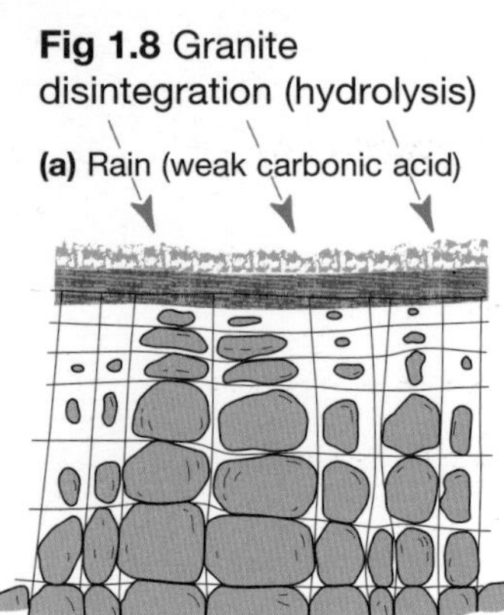

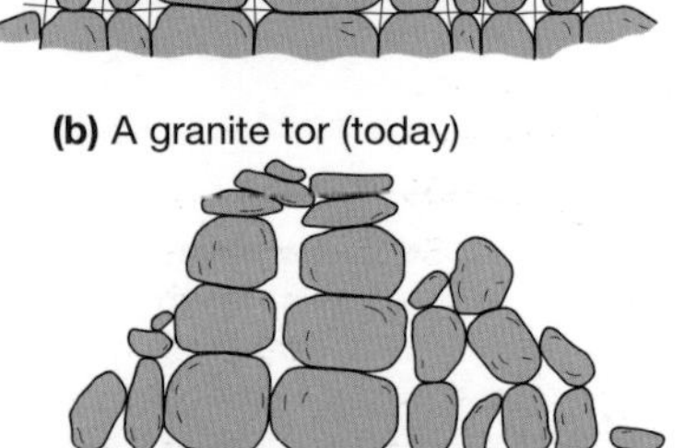

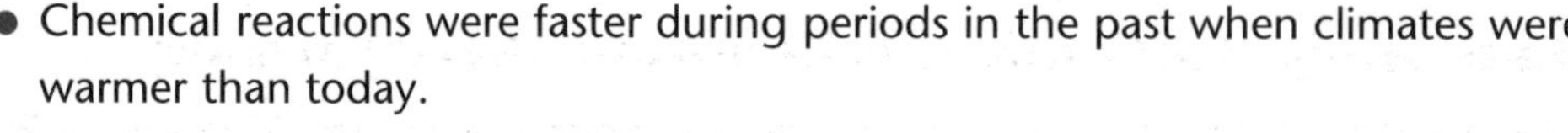

- Chemical reactions were faster during periods in the past when climates were warmer than today.
- Loose, disintegrated rock is left surrounding unweathered blocks of granite (Fig. 1.8a).
- If the weathered material is eroded and washed away, a pile of granite blocks will be left – a **tor** (Fig. 1.8b). On Dartmoor, tors are often found on higher points of the moor, e.g. Hounds Tor, Combestone Tor.

Biological weathering

The roots of plants and trees growing in joints and crevices on a rock face expand with age and force the joints apart.

Rotting vegetation creates humic acid, which adds to the processes of chemical weathering.

Landscapes – case studies

AQA A AQA C

Landforms in limestone areas

Surface landforms caused by solution:

- **limestone pavement** – beds of limestone that have been exposed to give flat areas of rock. Solution has widened the joints called grykes, while the remaining upstanding blocks are known as clints (Fig. 1.7).
- **swallow holes (potholes)** – enlarged joints which allow surface streams to take underground routes which follow bedding planes and joints in the limestone.
- **limestone gorges** – deep, steep-sided valleys often occupied by a stream, which may be the result of a cavern roof collapsing.
- **limestone scars** – exposed beds of horizontal limestone on valley sides, frequently with areas of scree below them due to freeze-thaw weathering.
- **resurgence** – places where streams re-emerge from an underground course, to flow on the surface again. Often this is at the junction of limestone with impermeable rock below.

Make sure you understand the process of limestone solution (page 13).

1. Locate the surface and underground landforms on the block diagram (Fig. 1.9, page 15).
2. Link their formation to chemical weathering (solution).
3. Answers will benefit from labelled diagrams.
4. Do some research to know located examples.

Underground landforms caused by solution and deposition:

- **caverns and caves** – are the result of widened bedding planes which have often been further enlarged by roof collapse, e.g. Ingleborough Cave System (Fig. 1.9).
- **stalactites and stalagmites** – water seeping and dripping from joints in the roof of a cavern, re-deposits calcium carbonate as features that grow down from the roof ('tites) and up from the floor ('mites).

Case study: Carboniferous limestone (karst) scenery

Fig 1.9 Block diagram of the limestone area around Ingleborough, Yorkshire Dales N.Park (based on OS 1:50 000 map no. 98)

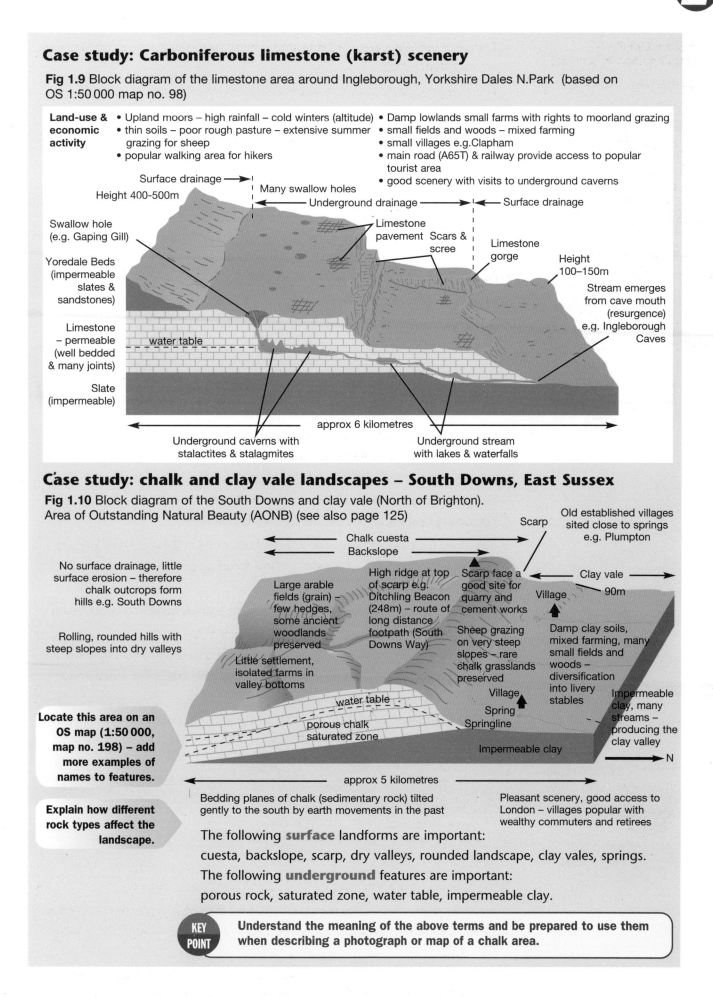

Land-use & economic activity

- Upland moors – high rainfall – cold winters (altitude)
- thin soils – poor rough pasture – extensive summer grazing for sheep
- popular walking area for hikers

- Damp lowlands small farms with rights to moorland grazing
- small fields and woods – mixed farming
- small villages e.g.Clapham
- main road (A65T) & railway provide access to popular tourist area
- good scenery with visits to underground caverns

Surface drainage
Height 400-500m
Many swallow holes
Underground drainage
Surface drainage
Swallow hole (e.g. Gaping Gill)
Limestone pavement
Scars & scree
Limestone gorge
Height 100–150m
Yoredale Beds (impermeable slates & sandstones)
Stream emerges from cave mouth (resurgence) e.g. Ingleborough Caves
Limestone – permeable (well bedded & many joints)
water table
Slate (impermeable)
approx 6 kilometres
Underground caverns with stalactites & stalagmites
Underground stream with lakes & waterfalls

Case study: chalk and clay vale landscapes – South Downs, East Sussex

Fig 1.10 Block diagram of the South Downs and clay vale (North of Brighton). Area of Outstanding Natural Beauty (AONB) (see also page 125)

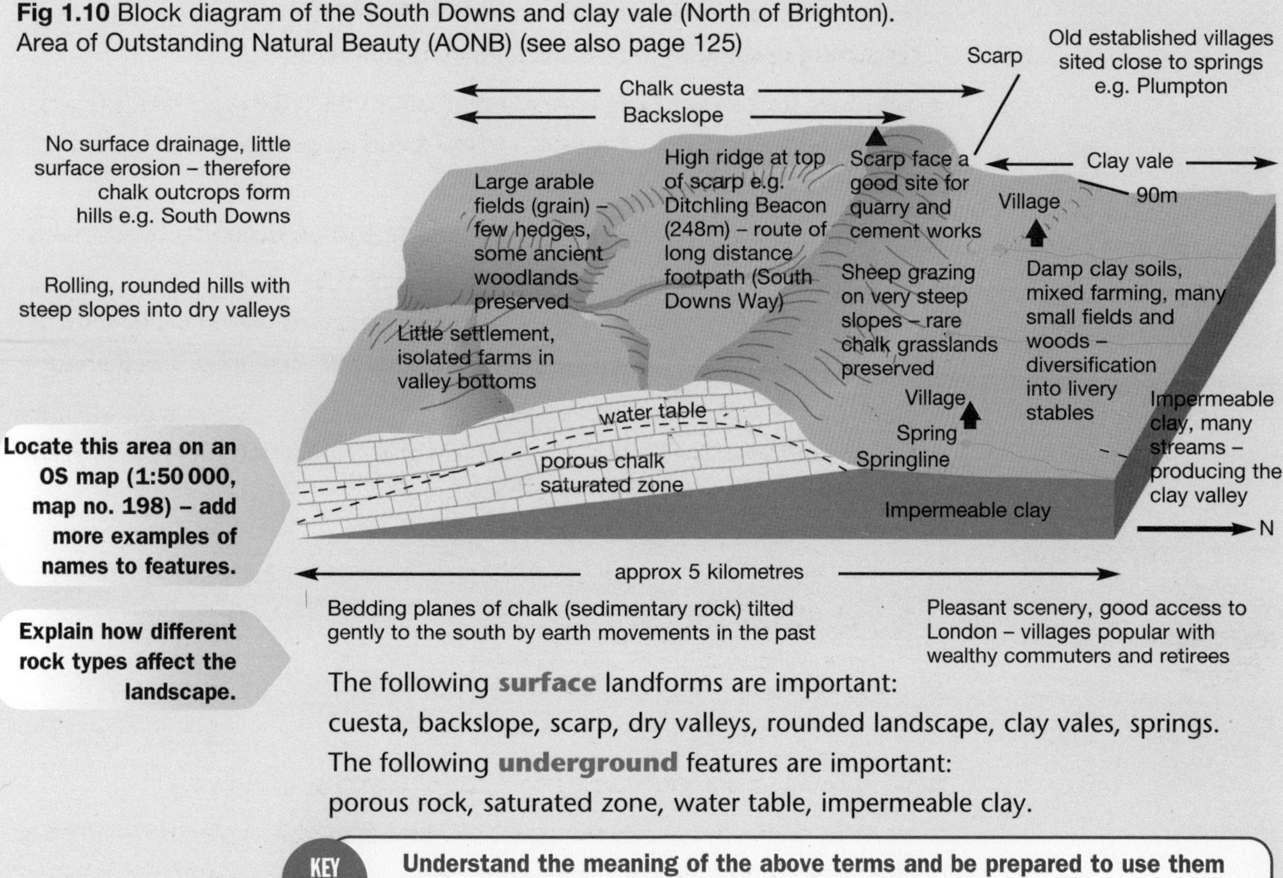

The following **surface** landforms are important:
cuesta, backslope, scarp, dry valleys, rounded landscape, clay vales, springs.
The following **underground** features are important:
porous rock, saturated zone, water table, impermeable clay.

KEY POINT **Understand the meaning of the above terms and be prepared to use them when describing a photograph or map of a chalk area.**

Case study: granite landscapes – Dartmoor, Devon (Dartmoor National Park)

Fig 1.11 Annotated map of the Dart Valley and surrounding moors (adapted from OS 1:50 000 map no. 191 & 202) (See also page 187)

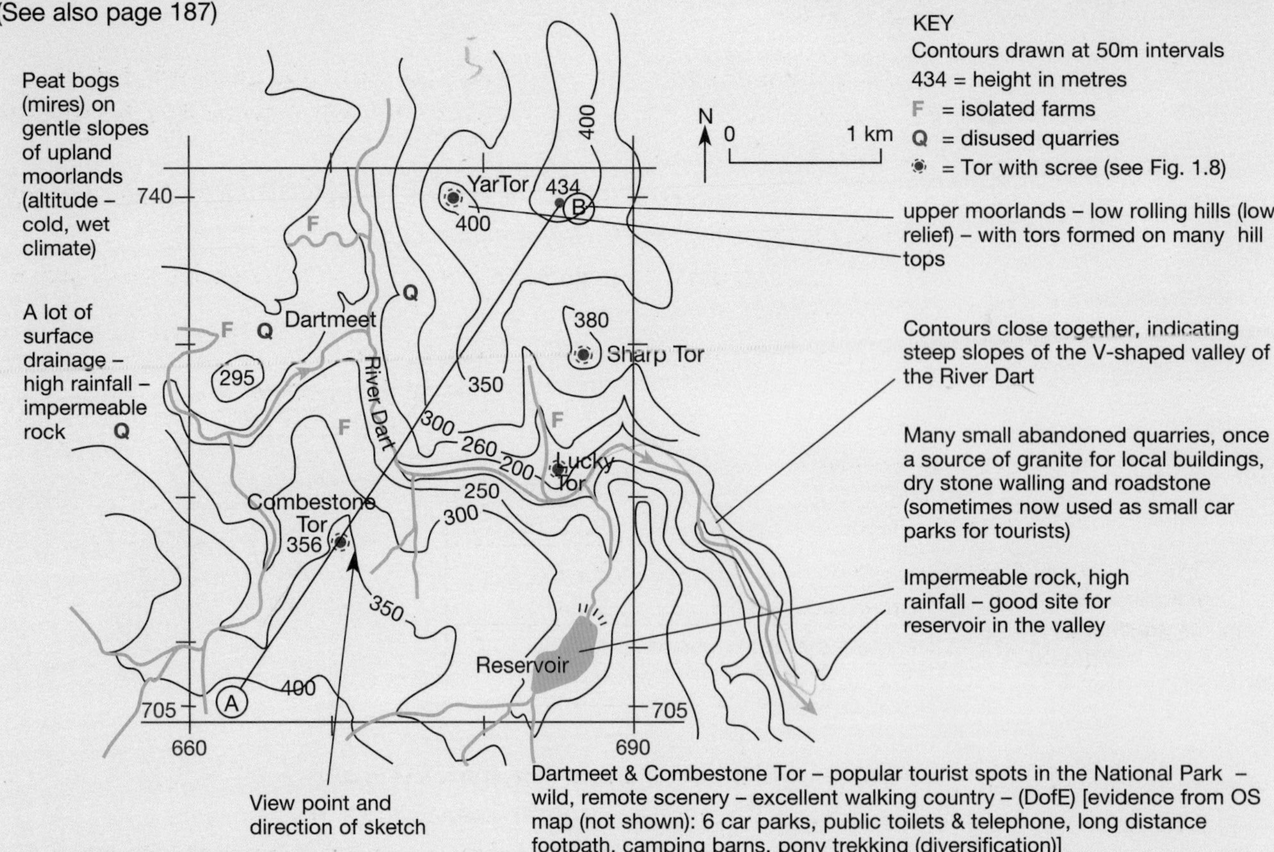

1. Draw a cross-section from A to B (marked on map). Use a vertical scale of 4mm: 50m
2. Label the different parts of the section using ideas from the map and field sketch.
3. Explain some of the features you have identified.

Fig 1.12 Field sketch of Combestone Tor and Dartmeet (see map above, Fig.1.11)

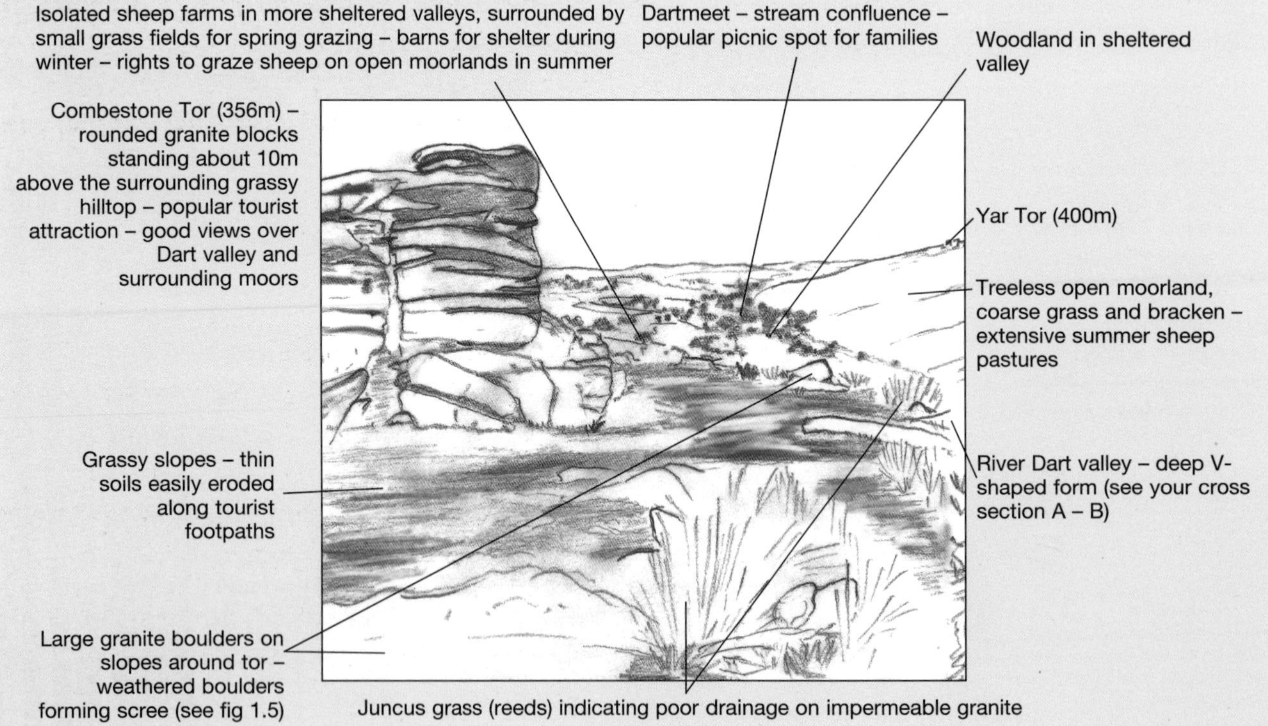

Quarrying as a contemporary issue

AQA A AQA C

Fact file

- Quarrying is a **primary** industry; there are over 1500 quarries in the UK. Surface extraction of coal and some other minerals is generally referred to as open-cast mining.
- **Aggregates** is a term used for all quarry products (e.g. limestone, granite) in whatever form (e.g. sand, gravel or crushed rock).
- The most important aggregates are limestone (includes chalk), sand, gravel and granite.
- Aggregates in the UK are used for the following:

Table 1.1

Use	% Aggregates
Roads	32%
Housing	25%
Other public works	16%
Factories & warehouses	13%
Offices & shops	14%

Choose a suitable method and draw a graph to show this data.

- Quarrying is an important part of the UK economy.
  - The sale of aggregates is worth over £3000 m per year.
  - Employment – about 30 000 people are directly employed with many others in linked industries (e.g. ready mixed concrete).
  - Local councils receive significant amounts of tax from the quarry companies.
- The government requires local councils to agree sufficient planning proposals to maintain an adequate supply of aggregate for the next ten years (the **landbank**).
- Proposals for new quarries or extensions to existing sites require planning permission from the local council. Public concern has increased over the years and proposals are often opposed by:
  - *national bodies* e.g. CPRE (Council for the Protection of Rural England) and Friends of the Earth
  - *local councils and residents*

Demand for aggregates

- The greatest demand is from the construction industry.
  - major projects such as the Channel Tunnel use up to 10 m tonnes
  - 125 000 tonnes are used in constructing 1km of a new motorway
  - building an average house uses 50–60 tonnes
- Demand doubled from 150m tonnes in 1975 to 300 m tonnes in 1998, some experts think it might double again in the next 20 years.

Describe the trend in demand shown on the graph (use figures).
Can you suggest reasons for the trend you have described?

Fig 1.13 Graph showing changing demand for aggregates 1970–2000

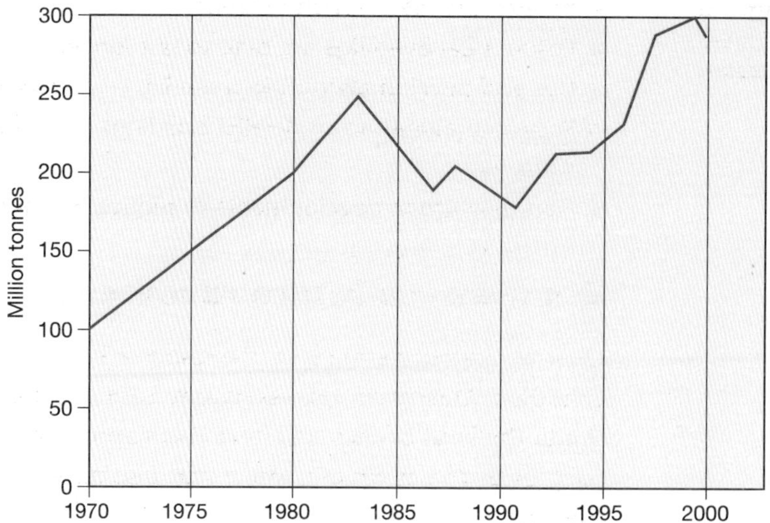

A. Arguments for quarrying

- provides essential products to the economy – limestone for cement and lime for fertiliser; clay for bricks; slate for roofing; sand and gravel for the construction industry
- local employment and income
- income for local councils (taxes)
- improved roads for access

B. Arguments against quarrying

- causes dust pollution from blasting operations, quarry machinery, crushing and sorting plants
- visual pollution from spoil heaps and large ugly buildings
- scenic pollution: ugly scars on hillsides can be seen from long distances
- noise pollution from blasting and lorry traffic
- heavy lorries cause vibrations and damage to buildings in villages; and damage and congestion along narrow country lanes

Fig 1.14 A cartoon used by CPRE to illustrate some of the issues of quarrying and its effects on the environment

Use the cartoon to describe and explain some of the issues connected with quarrying. Do you think there is any bias shown in the cartoon? Give a title to the cartoon.

Discuss the advantages and disadvantages of some of these strategies.

Some alternative strategies to reduce demand:

- Reduce the rate of road building.
- Ensure new buildings are built to last longer.
- Use and develop alternative materials (e.g. powdered glass for aggregate).
- Recycle materials from derelict buildings and when replacing old road surfaces.
- Plan new urban developments to reduce demand for aggregates.

Quarries – a future resource?

Quarries are not permanent, the resource becomes exhausted or too expensive to exploit. Quarry companies have to submit plans for the **restoration** of the site to the local council and have them approved. The costs of restoration are paid for by the company, which may also have to maintain the site for up to five years.

Can you link these different methods of restoration to examples in your region?

How successful have they been?

Are they sustainable?

Some examples of the use of former quarries include:

- *nature reserves and conservation sites* – natural regeneration under the management of the local Wildlife Trust or the RSPB can provide a range of habitats and preserve rare plants or important geological features
- *landscaping to accelerate natural regeneration* – contouring waste materials, reducing the height of rock faces, planting trees and shrubs
- *amenities in National Parks* – small quarries close to popular beauty spots may be used as car parks (see Fig 1.11 – Dartmoor)
- *sand and gravel pits can be restored to farmland*
- *landfill* – controlled management of waste disposal including household waste
- *urban infill* – when close to urban areas quarries can be developed for industrial estates, retail parks or housing estates; such use will reduce the demand for greenfield sites
- *recreation* – flooded quarries can provide sites for water sports such as sailing, windsurfing, fishing and jet skiing
- *water supplies* – large flooded quarries may combine this public facility with recreation.

Case study of a quarry

Use the internet or other resources to research a site where there are proposals to extend an existing quarry, or to open a new one.
Adapt the following ideas and headings to the proposal you have chosen.
You will need to:

Using the Fact File, list the main advantages and disadvantages of the quarry.

A. For a new or existing quarry establish a **Fact File** to cover:

Date started	
What it produces	
How much / year	
Cost to set up	
Planned life / reserves	
Methods of quarrying – (e.g. benches, blasting, crushing and sorting plant)	
Employment	
Method of transport (out of the quarry)	
Environmental protection measures (e.g. sound proofing, landscaping, working hours, blasting programme)	
Restoration plans	

B. If you have chosen an extension to an existing quarry, develop a similar fact file for the new proposals.

C. List the reasons given by the quarry company (a stakeholder) why their new proposals are needed – try to group them as **short term** or **long term**.

D. Who are opposing or might oppose the proposals?

List some stakeholders. These might include:

- local residents
- county, district and parish councils
- local wildlife groups
- CPRE, RSPB, Ramblers Association
- Friends of the Earth

List their **reasons** under the following headings:

Economic – Social – Environmental

E. Do you support the proposals or reject them?

a. Make a decision **giving your reasons**.

b. Which of the stakeholders you identified in D will support your decision? Give reasons.

c. Which other stakeholders might disagree? Give reasons.

Chapter 2

River landscapes and hydrology

After studying this section, you should understand:

- ***how a drainage basin functions as a system and is part of the hydrological cycle***
- ***that a storm hydrograph shows changes in discharge following a storm and that the shape of the graph depends on a range of factors***
- ***how rivers erode, transport and deposit material***
- ***that river processes vary between upland and lowland areas***
- ***that there are characteristic landforms associated with upland and lowland river valleys***
- ***the variety of factors that are responsible for flooding***
- ***that flooding is a complex issue to manage***

The drainage basin and the hydrological cycle

AQA A AQA B AQA C

The **drainage basin** is the area of land drained by a river and its tributaries. It is defined from other basins by its **watershed**.

Fig 2.1 The drainage basin of the River Adur, Sussex (based on OS 1:50 000 map no.198)

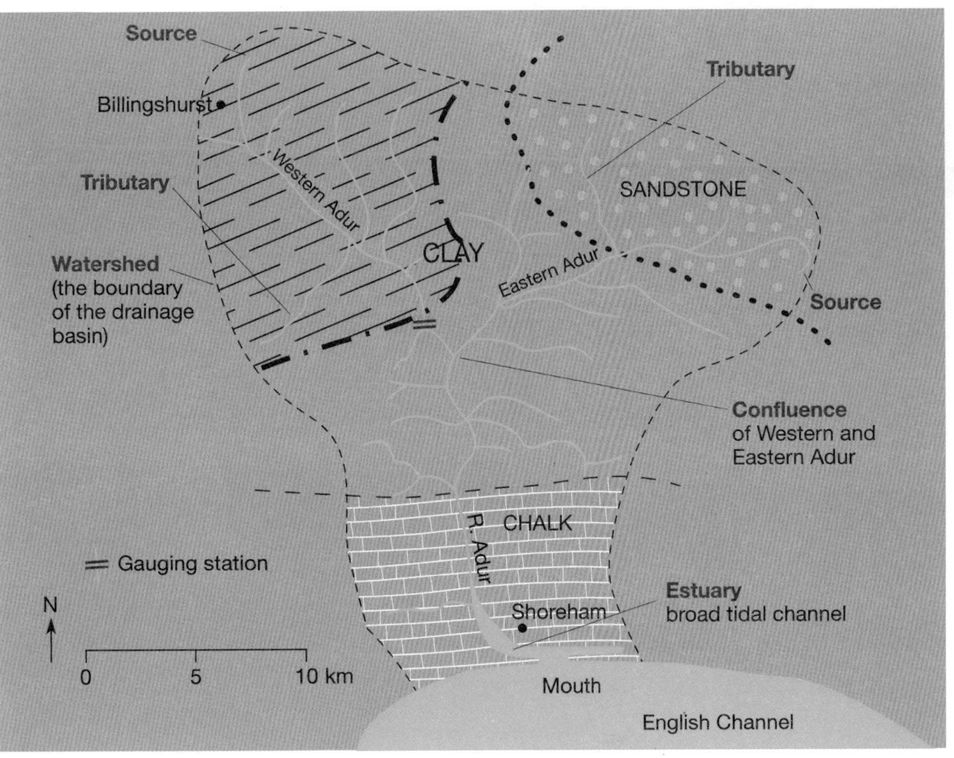

The **gauging station** measures the discharge of the river, which is the amount of surface water draining from the shaded area on the map

Large number of rivers on the impermeable clay (**high drainage density**)

Very few rivers on the porous chalk (**low drainage density**)

The drainage basin forms part of the hydrological cycle.

Fig 2.2 The hydrological cycle (a closed system – no water is gained or lost)

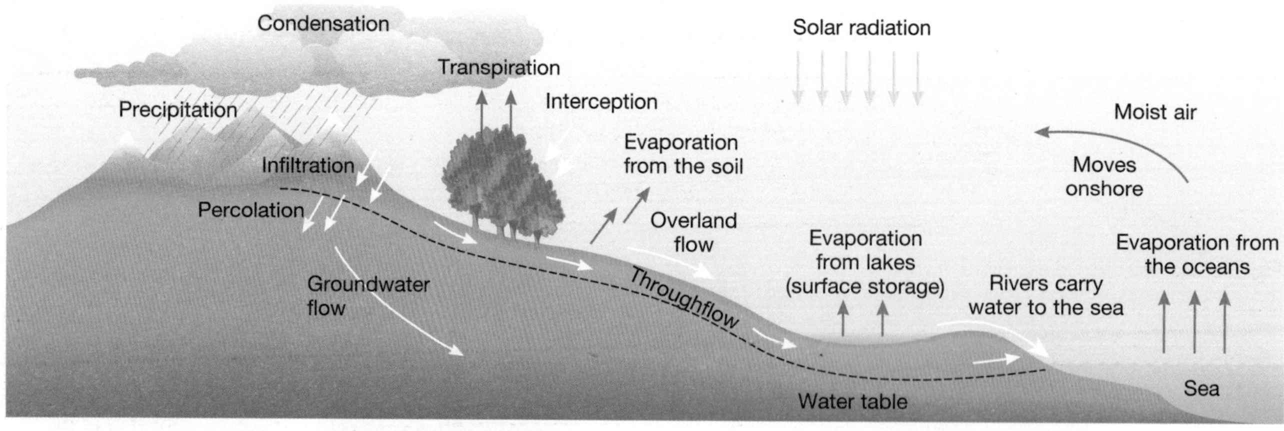

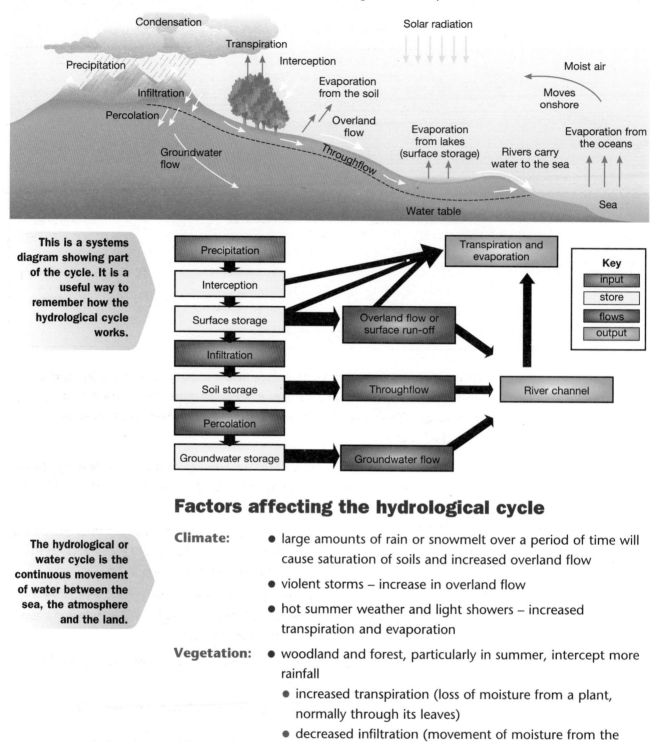

This is a systems diagram showing part of the cycle. It is a useful way to remember how the hydrological cycle works.

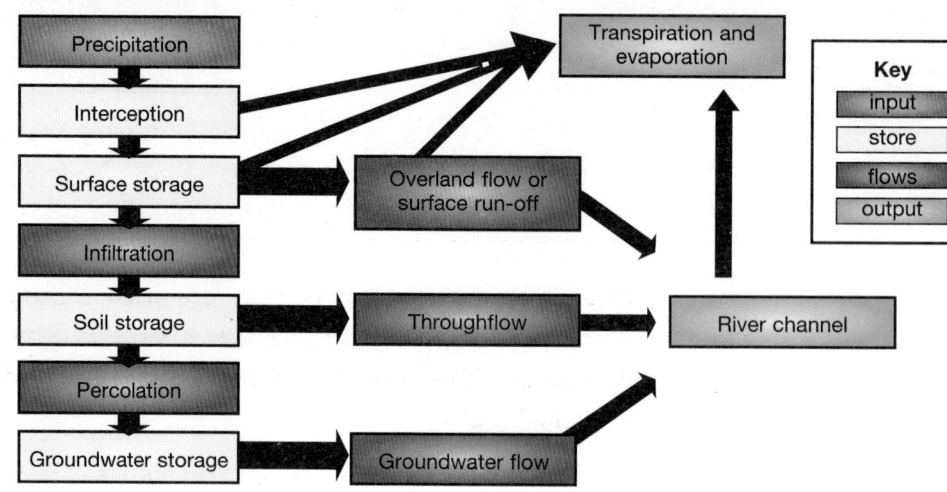

Factors affecting the hydrological cycle

The hydrological or water cycle is the continuous movement of water between the sea, the atmosphere and the land.

Climate:
- large amounts of rain or snowmelt over a period of time will cause saturation of soils and increased overland flow
- violent storms – increase in overland flow
- hot summer weather and light showers – increased transpiration and evaporation

Vegetation:
- woodland and forest, particularly in summer, intercept more rainfall
  - increased transpiration (loss of moisture from a plant, normally through its leaves)
  - decreased infiltration (movement of moisture from the surface into the soil or rock)

Slopes:
- steep slopes encourage overland flow and decreased infiltration

Soils:
- deep permeable soils increase infiltration and throughflow

Rock type:
- impermeable rocks increase overland flow
- porous rocks increase groundwater storage and flow

River discharge and the hydrograph

AQA A AQA B

The **discharge** of a river is **the amount or volume of water flowing past a particular point at any given time**.

KEY POINT

Discharge (Q) is the average velocity (V) of the river multiplied by the cross-sectional area (A) of the river.
$Q = V \times A$ = (m/sec x m²/sec) = m³/sec
(cumecs i.e. cubic metres per second)

Fig 2.3 Measuring river discharge

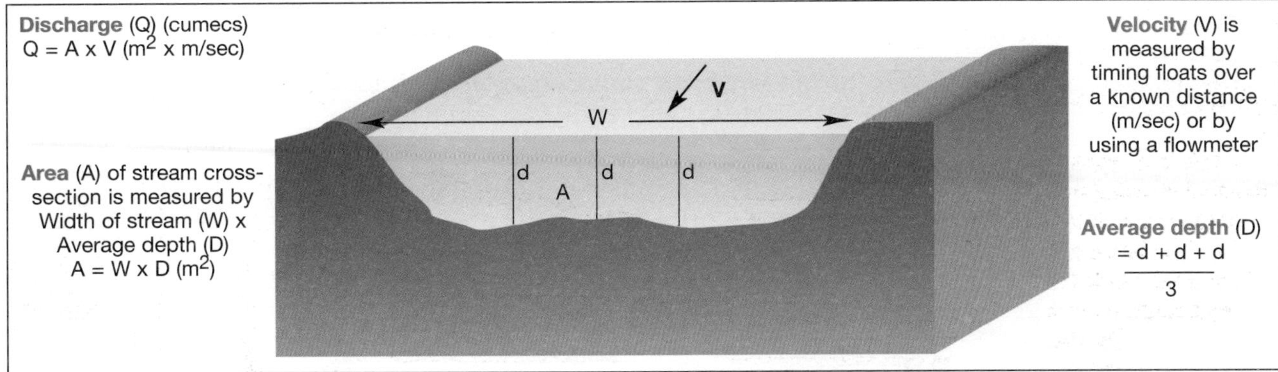

Official measurements of discharge are taken at gauging stations along the river. The changes in discharge over a period of time are recorded on a hydrograph. The **flood hydrograph** shows how a river responds to an individual storm.

Fig 2.4 The flood hydrograph

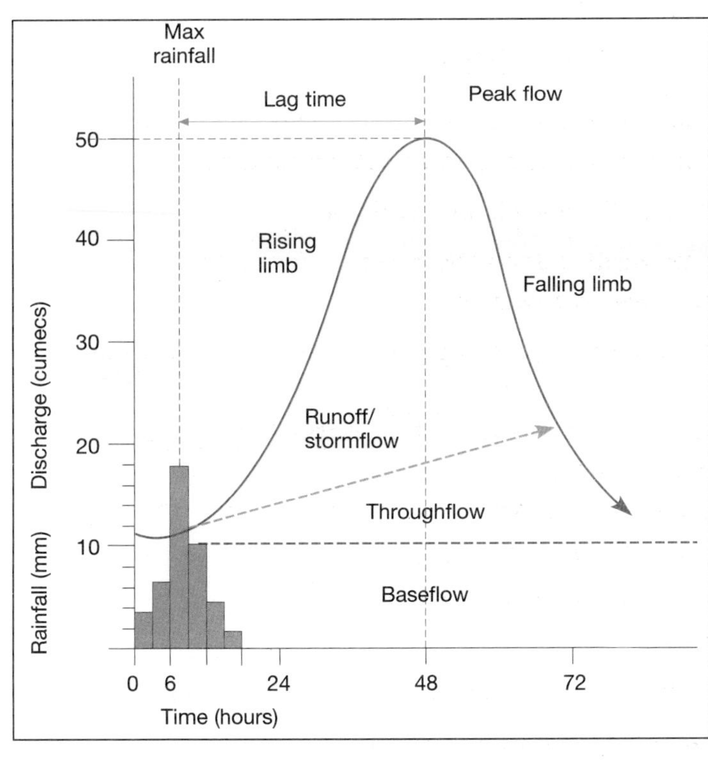

- A storm has been recorded in the drainage basin during the first 12 hours; maximum rainfall was at **6 hours**.
- The rising limb shows discharge slowly increasing as rainfall close to the gauging station finds its way into the river by overland flow.
- Discharge increases more rapidly after 24 hours, as rain which fell throughout the drainage basin flows down the river.
- A peak discharge of 50 cumecs is recorded at **48 hours**.
- The lag time is 42 hours (48 – 6 = 42).

KEY POINT

Lag time is the time difference between maximum rainfall and peak flow. Lag time is important; rivers with a short lag time are more likely to flood.

You may be given a storm hydrograph in an exam and asked to 'describe what happens to discharge' and 'work out the lagtime'.

Relate these factors to the inputs, stores and flows in Fig. 2.2.

This often appears in exam questions. Make sure you can explain why they will affect lag time. Use the correct terms from Fig. 2.2.

Factors affecting lag time and peak flow

Climate: rainfall; snowmelt

River basin: vegetation; slopes; soil; rock type

Human activity can decrease lag time by: draining more fields by underground drainage pipes; removing woodland and forests (deforestation); building more houses and roads (urbanisation); straightening river channels and artificially raising the height of river banks.

Human activity can increase lag time by: planting more woodlands and forest; building dams to store and control discharge; extracting water for industrial and domestic use; allowing flood water to be temporarily stored on the floodplain or in 'washes' during the winter.

A decrease in lag time can increase the likelihood of flooding.

River processes and landforms

AQA A  AQA B  AQA C

Rivers flow from their source to the sea. The slope the river follows is called the **long profile**.

Fig 2.5 The long profile of a river

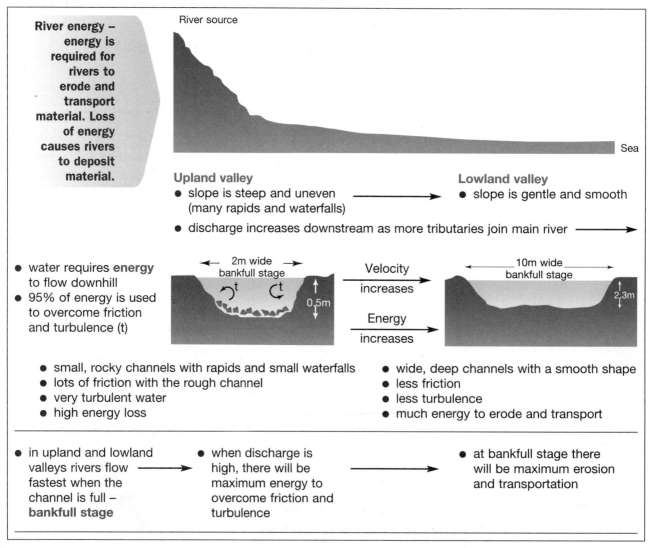

River energy – energy is required for rivers to erode and transport material. Loss of energy causes rivers to deposit material.

River source

Sea

Upland valley
- slope is steep and uneven (many rapids and waterfalls) →

Lowland valley
- slope is gentle and smooth

- discharge increases downstream as more tributaries join main river →

- water requires **energy** to flow downhill
- 95% of energy is used to overcome friction and turbulence (t)

2m wide bankfull stage; 0.5m

Velocity increases →

Energy increases →

10m wide bankfull stage; 2.3m

- small, rocky channels with rapids and small waterfalls
- lots of friction with the rough channel
- very turbulent water
- high energy loss

- wide, deep channels with a smooth shape
- less friction
- less turbulence
- much energy to erode and transport

- in upland and lowland valleys rivers flow fastest when the channel is full – **bankfull stage** →
- when discharge is high, there will be maximum energy to overcome friction and turbulence →
- at bankfull stage there will be maximum erosion and transportation

River processes

Erosion – wearing away the banks and bed of the river channel.

Transportation – the movement of eroded sediments (boulders, sand and mud) downstream.

Deposition – transported sediments are deposited by the river.

- Erosion and transportation are closely linked since the transported sediments greatly increase the ability of the river to erode.
- The amount of erosion and transportation is greatest when the river is at bankfull stage and there is most surplus energy.

Erosion – takes place by four processes:

Learn these terms carefully and use them when explaining how rivers erode.

- **abrasion or corrasion** – sediments carried along rub and abrade the channel sides and bed; pebbles swirled in cavities in the bed drill downwards forming circular **potholes**
- **attrition** – transported rocks collide, break up and become smaller and more rounded downstream
- **hydraulic action** – the immense weight and force of the flowing water removes material from the bed and sides of the channel
- **corrosion** – water reacts with minerals (chemicals) in the rocks which are carried away in solution. This is important where rivers pass over areas of limestone and chalk.

Transportation – sediments are transported in four ways:

- traction
- suspension
- saltation
- solution

A river's load is the amount of sediment being transported.

Fig 2.6 Processes of transportation

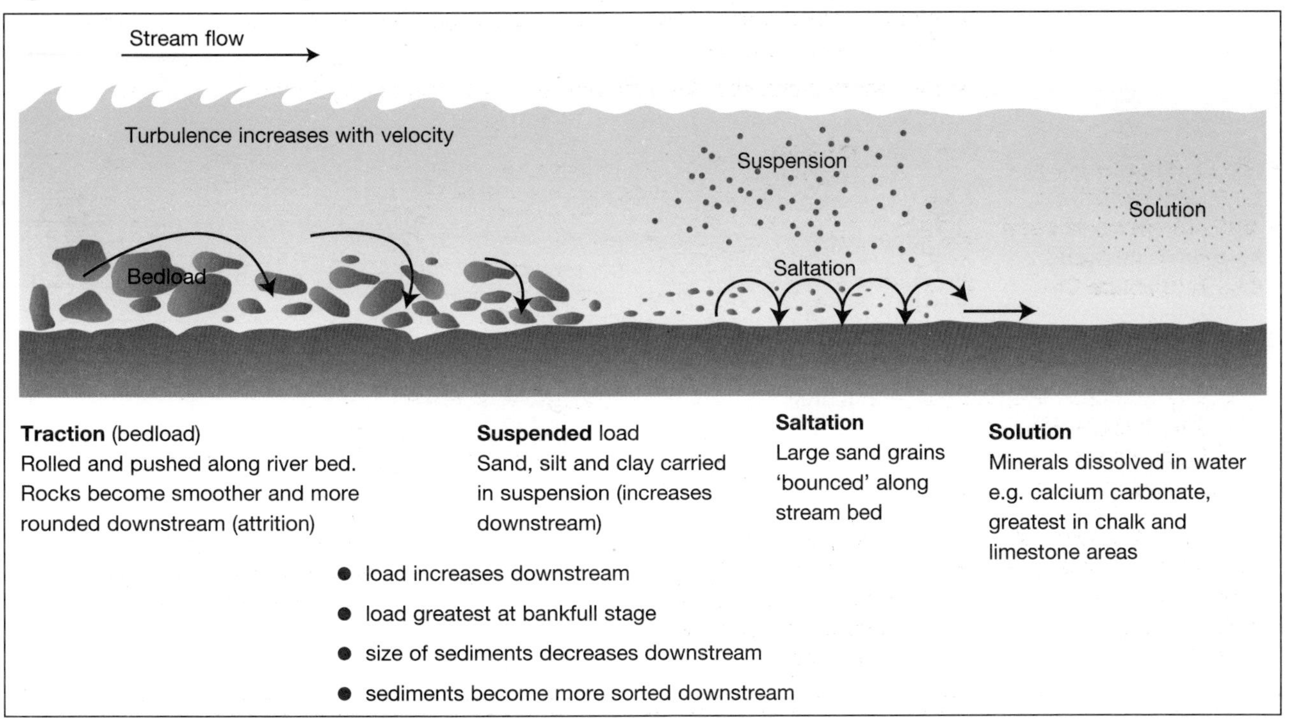

Stream flow
Turbulence increases with velocity
Suspension
Solution
Bedload
Saltation
Traction (bedload)
Rolled and pushed along river bed. Rocks become smoother and more rounded downstream (attrition)
Suspended load
Sand, silt and clay carried in suspension (increases downstream)
Saltation
Large sand grains 'bounced' along stream bed
Solution
Minerals dissolved in water e.g. calcium carbonate, greatest in chalk and limestone areas
load increases downstream
load greatest at bankfull stage
size of sediments decreases downstream
sediments become more sorted downstream

Deposition – occurs when a river lacks sufficient energy to transport the load it is carrying. Often this is when the river loses velocity:

- on entering a lake or the sea (e.g. an estuary – see Fig 2.1)
- on the inside of a bend or meander.

Fig 2.7 Rivers deposit their load on entering a lake (e.g. River Derwent entering Lake Bassenthwaite – Fig 4.11)

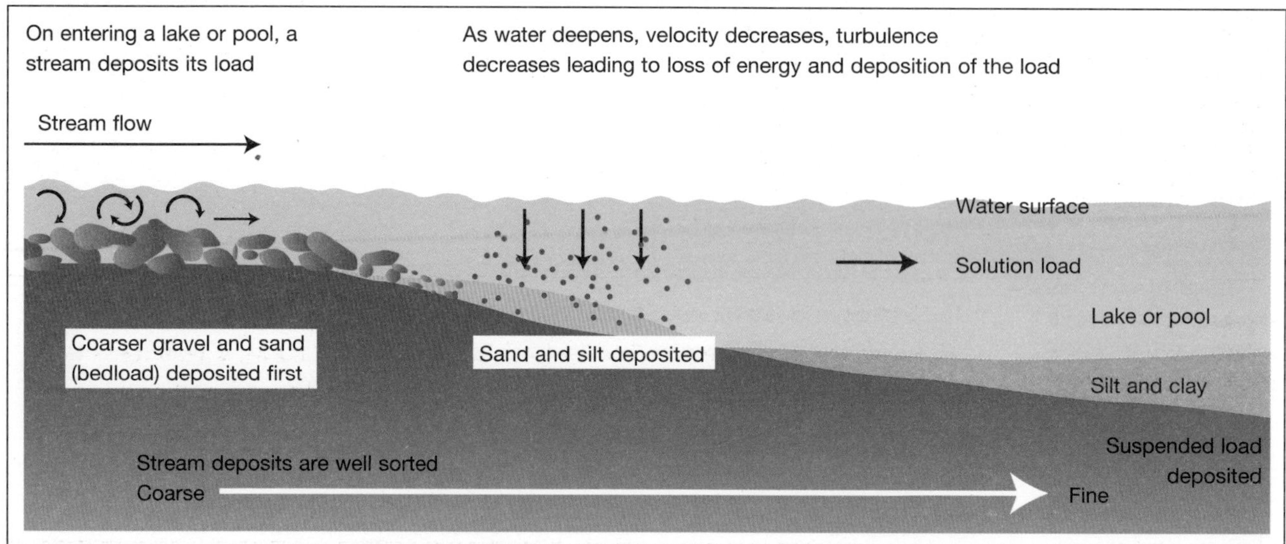
On entering a lake or pool, a stream deposits its load
As water deepens, velocity decreases, turbulence decreases leading to loss of energy and deposition of the load
Stream flow
Water surface
Solution load
Lake or pool
Coarser gravel and sand (bedload) deposited first
Sand and silt deposited
Silt and clay
Suspended load deposited
Stream deposits are well sorted
Coarse
Fine

River landforms

River landforms in an upland area

Many rivers have their source and headwaters in upland areas; the Rivers Severn and Dee rise in upland Wales.

Upland areas in the UK have:

- high precipitation with heavy snowfalls
- steep slopes
- impermeable rock
- thin soils and sparse vegetation
- areas where overland flow is very important.

Link these characteristics together to come to a better understanding of how the hydrological cycle works in upland areas.

Rivers in upland areas:

- have steep, uneven courses (long profile)
- have narrow, rocky channels and no flood plain
- have short lag times
- have rapid increases in discharge
- are very turbulent
- can transport boulders, rolling them along the bed by traction
- mainly erode downwards – **vertical erosion**.

Questions frequently ask for labelled diagrams explaining the formation of river landforms.

V-shaped valleys and interlocking spurs

E.g. the River Dart valley, Dartmoor (see page 16, Fig 1.11).

Fig 2.8 The formation of V-shaped valleys and interlocking spurs

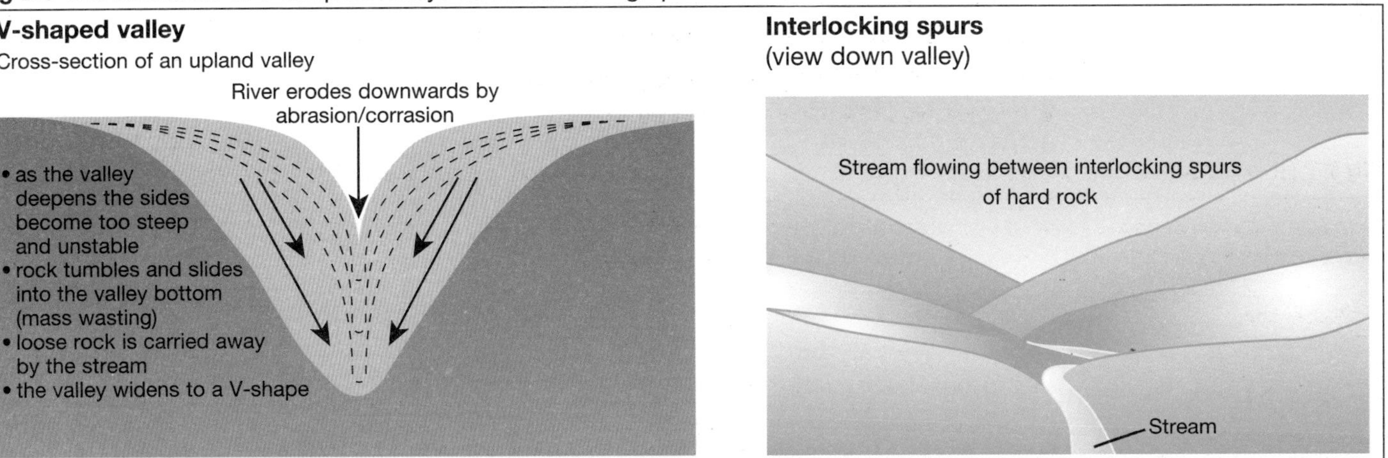
V-shaped valley
Cross-section of an upland valley
River erodes downwards by abrasion/corrasion
• as the valley deepens the sides become too steep and unstable
• rock tumbles and slides into the valley bottom (mass wasting)
• loose rock is carried away by the stream
• the valley widens to a V-shape
Interlocking spurs
(view down valley)
Stream flowing between interlocking spurs of hard rock
Stream

Remember to use and name real examples you have seen on field trips or holidays.

Waterfalls and gorges

Where rivers cross beds of harder (resistant) and softer rocks, a waterfall may form together with deep, steep-sided gorges, e.g. Cauldron Snout and High Force on the River Tees (N. Pennines) where the river crosses a hard band of rock called the 'Whin Sill' (see OS 1:50 000 map no. 91)

Fig 2.9 Formation of waterfalls and gorges

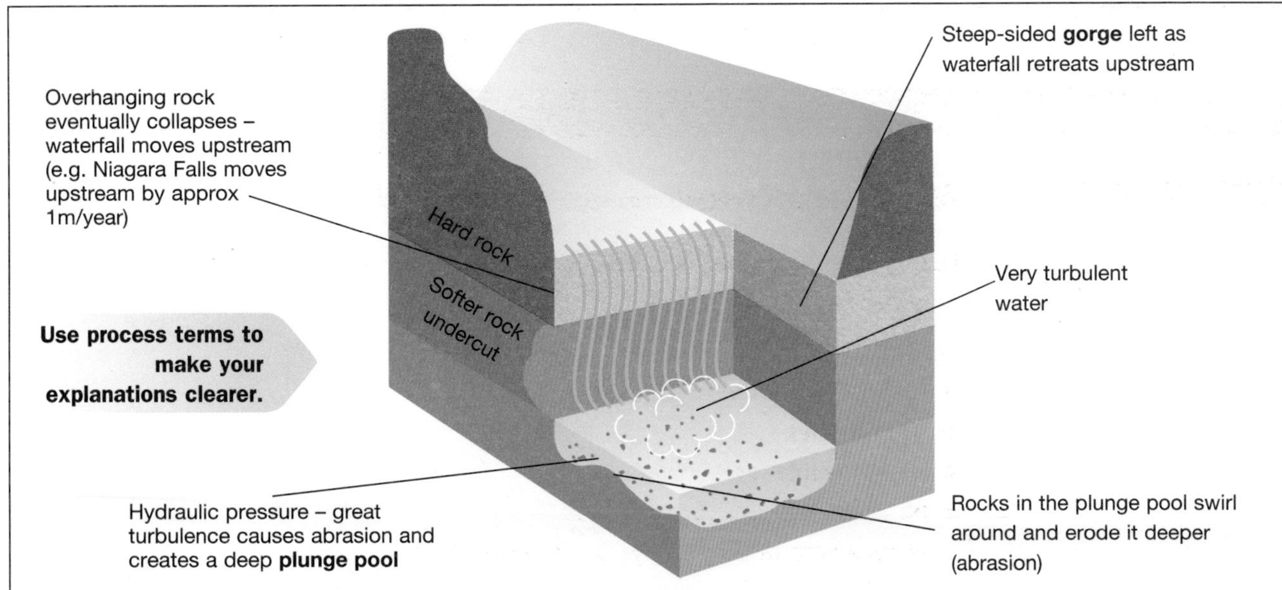
Steep-sided gorge left as waterfall retreats upstream
Overhanging rock eventually collapses – waterfall moves upstream (e.g. Niagara Falls moves upstream by approx 1m/year)
Hard rock
Softer rock undercut
Very turbulent water
Use process terms to make your explanations clearer.
Hydraulic pressure – great turbulence causes abrasion and creates a deep plunge pool
Rocks in the plunge pool swirl around and erode it deeper (abrasion)

Waterfalls may also be the result of:

- glacial erosion (see page 49, Fig 4.3a and 4.3b)
- a fall in sea level.
- Earth movements

River landforms in a lowland area

Rivers such as the Severn and Dee flow through wide, lowland valleys before reaching the sea.

Lowland areas in the UK, compared with the uplands, have:

- lower precipitation
- gentler slopes
- less resistant rocks
- deeper soils and more vegetation
- areas where throughflow, groundwater flow and storage are more important than overland flow.

Rivers in lowland areas:

- have low gradients, gentle long profiles
- have wide, deep, smooth channels
- have longer lag times
- have larger discharges (from the whole drainage basin) particularly at bankfull stage
- have stronger currents (velocity)
- have lower friction and turbulence
- have increased erosion and transportation
- have erosion which is mostly side-to-side – **lateral erosion**
- transport large 'loads' of fine sediment by saltation and suspension.

Meanders, floodplains and oxbow lakes

A common exam question asks you to: 'With the aid of diagrams explain the formation of an oxbow lake'.

With large discharges, strong currents (velocity), low friction and less turbulence, rivers in lowland areas have surplus energy to erode and transport. Lateral erosion erodes the banks of the channel and large bends or **meanders** develop. Since erosion occurs mostly on the outside of bends, where velocity is greatest, meanders increase in width. The valley sides are slowly eroded to create wide, flat **floodplains**. Meanders can develop extreme loops across the flooplain. Further erosion can leave 'loops' abandoned as **oxbow lakes**, when the river assumes a straighter, more **efficient** course.

Fig 2.10 Formation of meanders, flood plains and oxbow lakes

Valley cross-sections

Flood plain

Stage I
The fastest current moves from side to side within the channel

High energy

Bank erosion by hydraulic action and abrasion

Least energy (deposition)

Erosion of valley sides to create a floodplain

Stage II
Meanders develop into large 'loops'

Path of fastest flow

River cliff (erosion)

Slip-off slope (deposition)

Cross-section below

A

B

Narrow meander neck

Meander loop

Stage III
River eventually breaks through the neck of the meander, taking a straighter, more efficient course

Meander loop is abandoned as an **oxbow lake** on the flood plain (see Fig. 2.11)

Cross-section of stream channel (assymetric)

(A) River cliff

Slip-off slope (B)

Erosion of channel bank by hydraulic force and abrasion

Fastest flow

Slower flow

Deposition of some 'load' (finer sediment e.g sand)

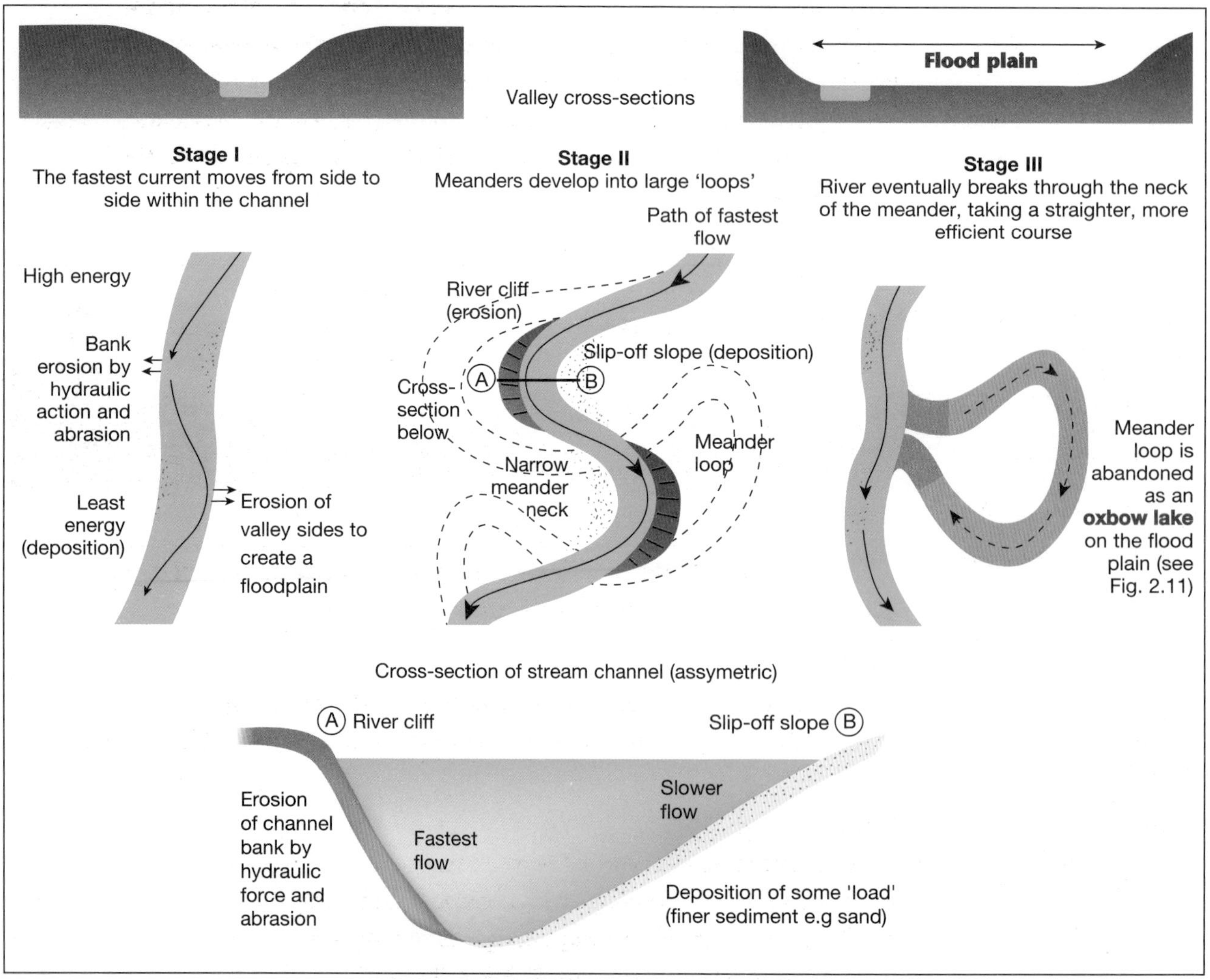

You may know local rivers with these features. Name them to support your answer, particularly if the question says 'using examples you have studied'.

Fig 2.11 Map of River Dee showing floodplain, meanders and oxbow lakes (based on OS 1:50 000 map no. 117)

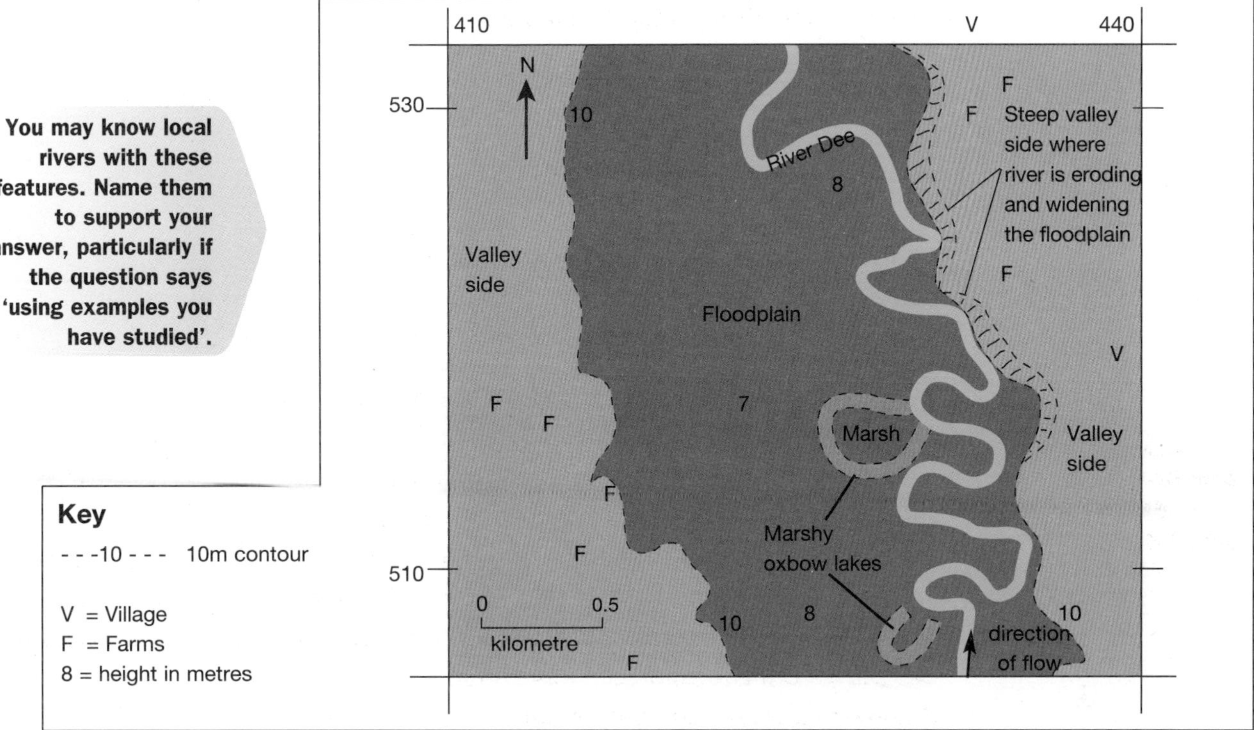

At times of very high discharge, rivers may overflow their banks and spread out across the floodplain. Floodplains are natural 'stores' of water until the floodwaters go down. At bankfull stage the river carries maximum 'load'. On flooding, the water slows suddenly, energy is lost, and deposition takes place. Coarse sediments (sand) are deposited immediately on the river banks, to form natural **levées**. Finer sediment (silt and clay) forms deposits across the floodplain called **alluvium**.

River Rhone, Switzerland – high levées have been constructed to protect the flood plain from flash floods – see pages 100–101 Figs 7.10 and 7.11.

Fig 2.12 Floodplains and levées

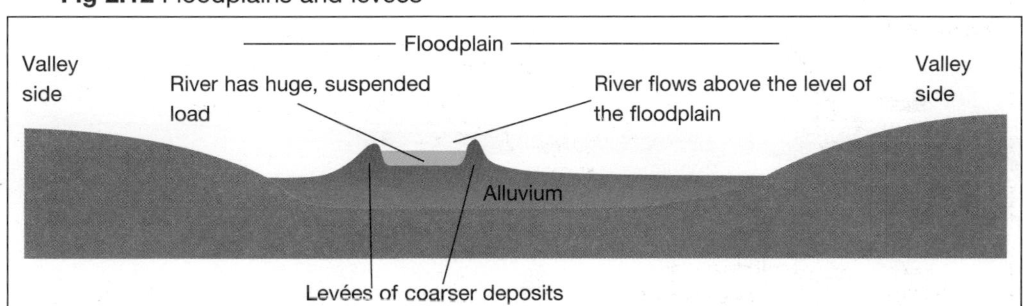

Alluvium is very fertile. In the lower Nile valley in Egypt, farmers rely on annual floods to bring fresh supplies of silt to their land.

Deltas

Suggest reasons why, in many LEDCs, there are dense populations of people living on deltas. Why might these be dangerous places to live?

Deltas form at the mouths of many of the world's larger rivers, e.g. the Nile (Egypt), the Ganges (Bangladesh), the Mississippi (USA) and the Mekong (Vietnam).

Deltas form where:

- rivers carry huge 'loads' of fine sediments
- currents slow on entering the sea cause deposition
- tidal currents are not strong enough to remove the sediment
- rivers enter lakes forming small deltas (Fig 2.7).

The river is broken up into many channels, called **distributaries**, which spread the sediment out into distinctive shapes.

Types of delta:

- arcuate, e.g. Mekong and Nile
- bird's foot, e.g. Mississippi

Fig 2.13 Types of delta

Nile delta – arcuate

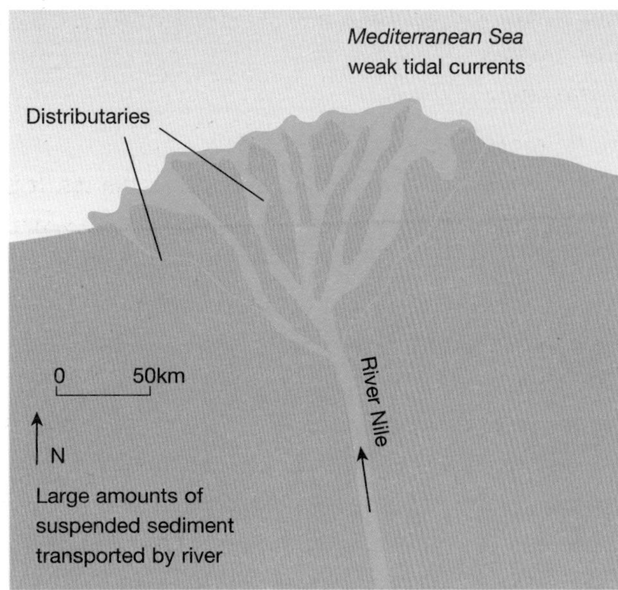

Mississippi delta – bird's foot

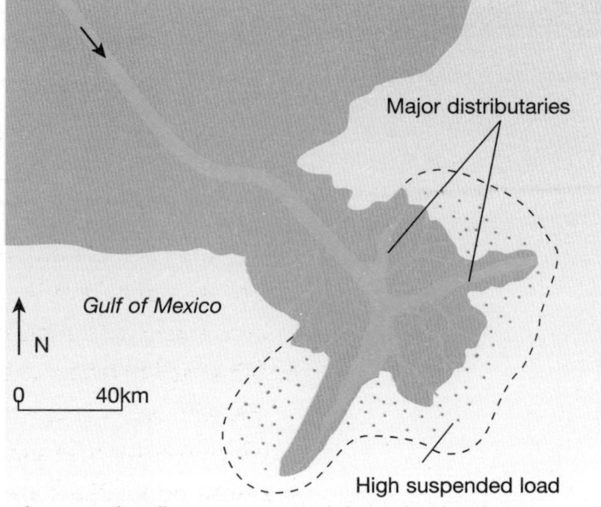

Questions often link estuaries to the location of large ports (e.g. Humber/Hull; Severn/Avonmouth) and to industries such as oil refineries (Dee/Connah's Quay). Explain why they are located there.

Estuaries

There are no large deltas around the coasts of the UK. Many rivers have estuaries, which are broad river valleys flooded daily by the tide. At low tide, large expanses of mud, sandflats and saltmarshes are exposed, which are mostly composed of sediments deposited by the river, e.g. Humber, Severn, Dee (see OS 1:50 000 map no.117).

River flooding and management

Flooding occurs when the discharge of a river is too great for its channel to hold. Water will flow over the banks and occupy the floodplain.

Floods can have huge social and economic costs, but they can also be beneficial. Vast numbers of people around the world live in areas that suffer from regular flooding.

Causes of flooding

Most flooding is a combination of climatic, drainage basin and human factors.

Research the causes of the Boscastle flood (N. Devon) in August 2004.

Climate factors:
- heavy rainfall is less likely to infiltrate into the soil and will reach the river quickly
- rapid snowmelt contributed to the 1995 Rhine floods

Drainage basin factors:
- impermeable rock, thin soils, steep slopes and saturated ground

Human activity factors:
- deforestation reduces interception and transpiration
- urbanisation and modern farming practices increase run-off

Impacts of flooding

- loss of life – people evacuated from their homes
- buildings and property damaged – high insurance costs
- crops ruined, animals drowned
- transport and communications flooded
- economic impact on shops, offices and industry, and public utilities
- sewage contaminates water supplies
- political pressures on government

Impacts in LEDCs are likely to be more severe than in MEDCs.

- Governments cannot afford to invest in costly flood prevention schemes. It is estimated that schemes in Bangladesh will cost over £1 billion over 30 years (Bangladesh already has a large international debt).
- Governments give higher priority to investing in industry and exports, to raise their GNP, than in flood prevention schemes.
- Schemes funded by the World Bank provide only long-term solutions – aid agencies are only able to support schemes to improve farming, schools, hospitals which will improve survival rates rather than schemes to prevent flooding.
- Dense populations are attracted to the fertile soils on floodplains and deltas where subsistence agriculture (see page 154) provides them with food (e.g. Mekong Delta) – these areas are likely to suffer annual floods.
- Living at subsistence level, a flood can destroy crops and livestock, causing food shortages, starvation and disease.
- Poor communications and medical services mean help is slow to arrive.
- Flood warning systems not well developed, many people are without radio and television.

Flood control

A. Hard strategies

- **Dams** can store and control river discharge (dams are used to control tributaries of the Mississippi, e.g. River Missouri).
- **Levées and retaining walls** – by raising the height of river banks, water can be contained. (November 2000 – flood walls in York contained record river levels and saved many homes from flooding.)
- **Straightening meanders** – increases speed of flow and reduces the length of the river (the length of the Mississippi has been shortened by over 240km).
- **Flood relief channels** – provide additional channel alongside existing course of river (the Maidenhead, Windsor and Eton Relief Channel, has removed 5500 homes from the threat of Thames floodwater).

Failure of dams or levées can cause very severe flooding (e.g. New Orleans 2004).

B. Soft strategies

- **Washlands or spillways** – in times of high discharge sluice-gates are opened and water is allowed to flood adjacent areas (washlands have been in use on the Great Ouse in Cambridgeshire for many centuries).
- **Afforestation** (the planting of woods and forests) – increases interception and reduces run-off (e.g. Tennessee river valley, a tributary of the Mississippi).
- **Planning regulations** – reject development proposals on flood plains so that they can still be allowed to flood (between 1970 and 1990 in the UK, two million homes were built on floodplains).

Where rivers cross international boundaries, river management and flood control are more difficult (e.g. Mekong River passes through six countries).

- Flood warning systems – In September 2000, the Environment Agency (UK) issued new flood warning codes to be broadcast on the radio and television, and produced flood plan checklists.

MEDC case study: Mississippi floods, August 1993

Causes

- There had been continuous rainfall from April to July; soils were saturated with rapid run-off.
- Heavy thunderstorms in June throughout the upper drainage basins of the Missouri and Mississippi gave over twice the normal rainfall.
- Huge drainage basins (30% of the USA); large rivers meeting at confluences, e.g. St. Louis, combined to give record discharges.
- Silt had raised the bed of the channel; levées had been raised. The river flowed above the level of the surrounding floodplain.
- Record discharges caused levées to collapse and the river eroded huge gaps in the banks and flooded the low-lying floodplains.
- Growth of large urban populations demanded higher flood defences and caused rapid run-off.

Questions frequently ask for case studies in an MEDC or LEDC. Make sure you can relate your answers to the place you have named.

Fig 2.14 The drainage basin of the Mississippi

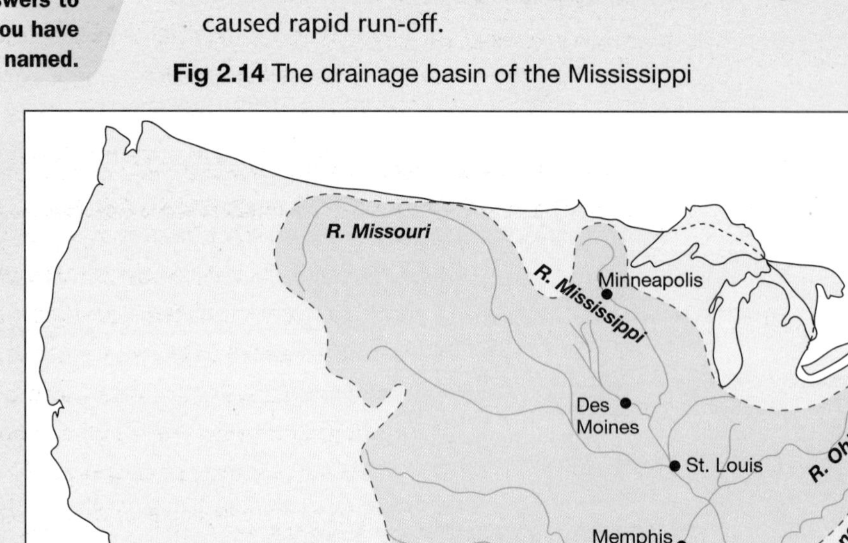

Impacts

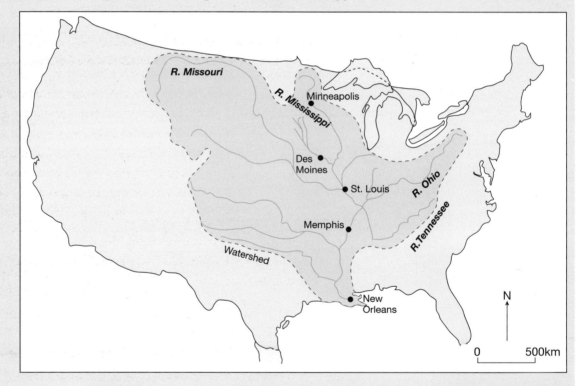

- Severe flooding occurred from Minneapolis to Memphis (8 million hectares).
- 47 people died, 74 000 were evacuated and 45 000 homes were affected.
- Des Moines and St. Louis lost electricity and water supplies.
- Sewage contaminated water supplies over large areas.
- Crops of grain and soya bean, valued at $6.5 billion, were ruined.
- Communications disrupted.
- Total cost estimated at $11 billion.

Raising levées creates greater potential dangers in the long-term, e.g. flooding in New Orleans 2004.

Outcomes

- Government declared a state of emergency; military aid supplemented local emergency services; states received large amounts of federal aid.
- 'Hard' defences soon repaired and raised to hold more water in the river – more spillways and washes have been designated to store surplus water in the short term.
- Most people had insurance policies to meet the costs of destruction.

LEDC case study: Mekong Delta (Vietnam) floods, 2000

Fig 2.15 The drainage basin of the Mekong river

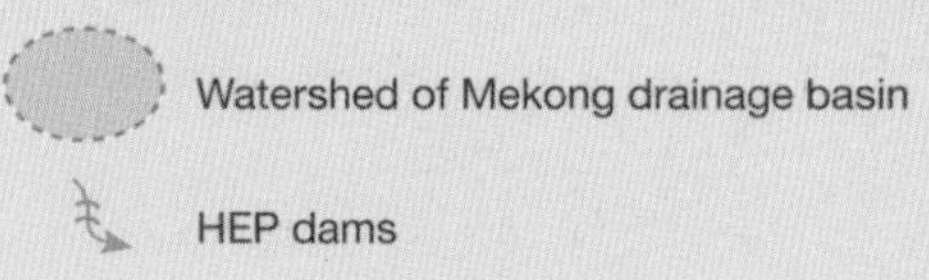

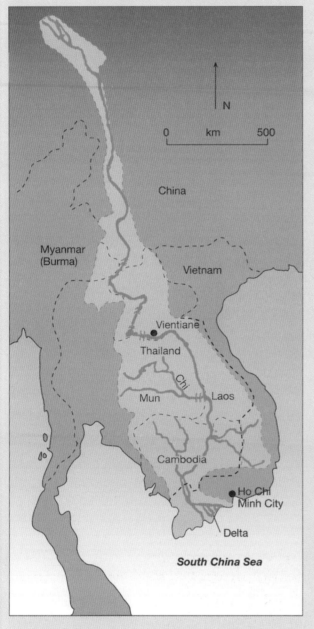

Record flood levels covering large areas of the delta and extending almost to the capital city, Ho Chi Minh.

Fact file: Mekong River

- length 4200 km
- source in China 5000 m above sea level
- Delta into China Sea
- HEP potential high
- 'Lifeblood' for up to 50 m people
  - fish supplies
  - irrigation of rice fields and orchards
  - important waterway

Causes

- The mountainous, steep-sided valleys in Laos and Cambodia received heavy rainfall.
- The monsoon rainfall (wet and dry seasons) was above average for the previous three years, so the ground was very saturated from flooding.
- Much of the rainfall fell in very intense storms.
- Rural communities in Laos and Cambodia depend on the river and forest for building materials, food and medicines; many are shifting cultivators. Population growth is causing deforestation.
- Poverty leads to illegal logging – 39 000 recorded violations of forestry regulations in one year.
- Concrete-lined riverbanks are increasing in urban areas such as Vientiane, the capital of Laos.
- Dam building in Laos and China is causing deforestation.
- Dams in China and Laos released large discharges of water without regard for the delta areas in Vietnam.
- Dense rural population living in the delta is causing overgrazing and soil erosion.

Explain how these factors might affect discharge and lag times.

Make a simple table to compare impacts and outcomes with an MEDC case study, e.g. Mississippi page 32

Impacts

- 350 people dead; mostly young children.
- Many hundreds of thousands forced to leave their homes in June to occupy small muddy banks. They were unable to return to their villages until December.
- Three successive years of flooding left no reserve food supplies; people had to rely entirely on relief workers to bring food.
- New high yielding rice strains are unable to withstand long floods; the year's rice crop was lost.
- A plague of golden water snails threatened all surviving crops.
- There were life-threatening diseases from contaminated water.

Outcomes

- The government has extended small loans so that families can build houses on stilts. It has also distributed thousands of small boats so that families can continue to fish for food during floods.
- The government has implemented a successful policy restricting families to two children only; population growth reduced.
- Dams in Laos will help control floods but will also reduce the amount of silt carried down by the river. Delta farmers rely on fresh fertile silt to replenish their land. The lack of silt will cause the delta to shrink in size and become lower – threatening worse flooding.

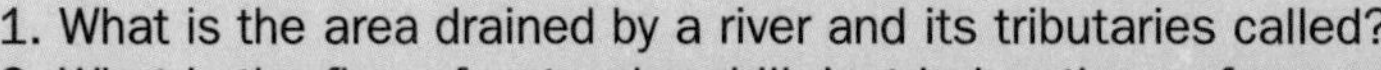

1. What is the area drained by a river and its tributaries called?
2. What is the flow of water downhill, just below the surface, called?
3. What are the units of discharge measured in?
4. On a hydrograph, what is the name given to the time difference between maximum rainfall and peak flow?
5. What is the name given to the movement of sediment along a streambed by short hops or jumps?
6. What shape would suggest to you that an upland valley had been eroded by a river?
7. What is the name given to an abandoned 'loop' of a river found on a floodplain?
8. What name is given to wide river valleys flooded daily by the tide such as the Humber?

1. Drainage basin 2. Throughflow 3. Cumecs (c^3m/sec) 4. Lag time 5. Saltation
6. V-shape 7. Oxbow 8. Estuaries

Chapter

3 Coastal landscapes

LEARNING SUMMARY

After studying this section you should understand:

- ***how waves form and the different types of wave***
- ***the processes of erosion, transportation and deposition***
- ***the landforms associated with erosion and deposition***
- ***how coastal defences can be managed to prevent flooding***
- ***that tourism and other activities along a coast have to be managed***

KEY POINT

The coast is the interface or narrow zone where the sea, land and atmosphere meet.

Coasts are shaped by the processes of erosion, transportation and deposition. The forces most responsible for these processes are:

- **waves**
- **tides**
- **currents.**

Coasts can change shape rapidly, as in a single storm, but most landforms are the result of long-term changes. The nature of the rocks forming the coast affects its shape and form.

Formation of waves

AQA A AQA C

- Waves result from the wind blowing over the surface of the sea. (This is not the case for tsunamis; see page 109.)
- The largest waves have the most energy.
- Wave size depends on:
  - the strength of the wind
  - the duration of the wind
  - the fetch.

Fig 3.1 The formation of waves

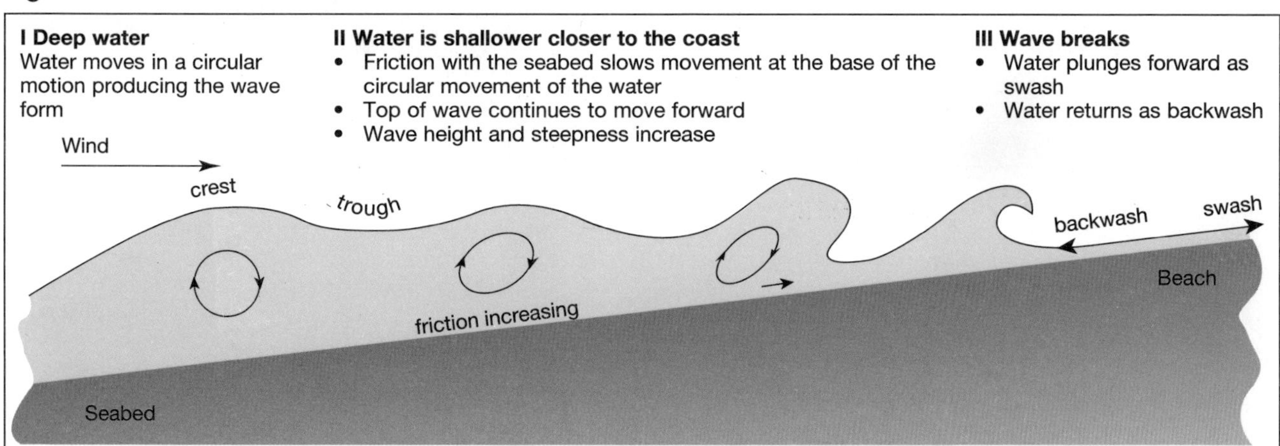

I Deep water
Water moves in a circular motion producing the wave form
II Water is shallower closer to the coast
• Friction with the seabed slows movement at the base of the circular movement of the water
• Top of wave continues to move forward
• Wave height and steepness increase
III Wave breaks
• Water plunges forward as swash
• Water returns as backwash
Wind
crest
trough
backwash
swash
Beach
friction increasing
Seabed

KEY POINT **The fetch is the distance the wind has travelled over the sea before reaching the coast.**

At Land's End, Cornwall, waves with the highest energy come from the south-west (Atlantic Ocean) because:

- frequent storms over the Atlantic give gale force winds
- winds blowing from the south-west are the **prevailing winds**
- winds from the south-west have a fetch of many thousands of miles.

Fig 3.2 Types of waves

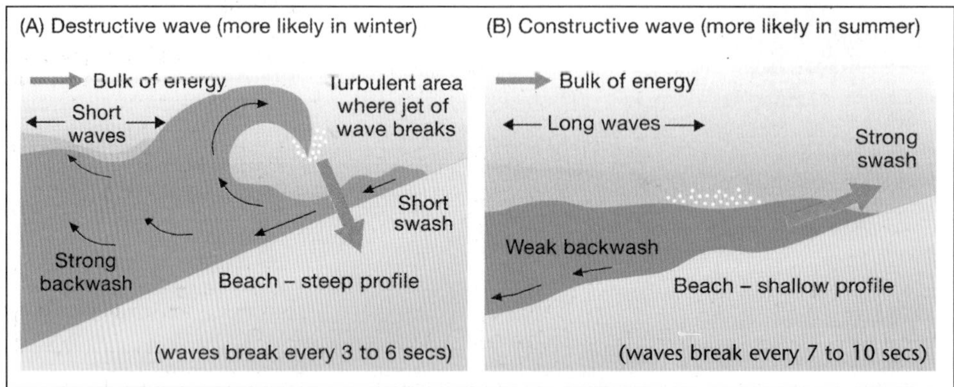

Tides

At high tide the water will be deepest offshore and larger waves with more energy can reach the beach or cliff. Storm waves at high tide have the highest energy of all and they are responsible for most erosion and transportation.

Processes of erosion

 AQA A  AQA C

Waves can erode in four different ways:

- **hydraulic action:** the pressure of the wave hitting a cliff traps and compresses air into joints and cracks, weakening the rock and breaking it up
- **abrasion (corrasion):** waves hurl rocks against the cliff and swirl sand backwards and forwards, wearing the rock away
- **attrition:** rocks and sand rub against each other, making them smaller and more rounded
- **corrosion (solution):** salt water corrodes minerals in many types of rock, causing rocks to break up.

Note the similar terms to rivers.

Weathering processes such as freeze-thaw (see page 12) break up the rock on the cliff face.

Mass movement: cliffs collapse when the waves erode and undercut the base of the cliff.

Landforms of erosion

AQA A AQA C

Headlands and bays

Fig 3.3 Formation of headlands and bays

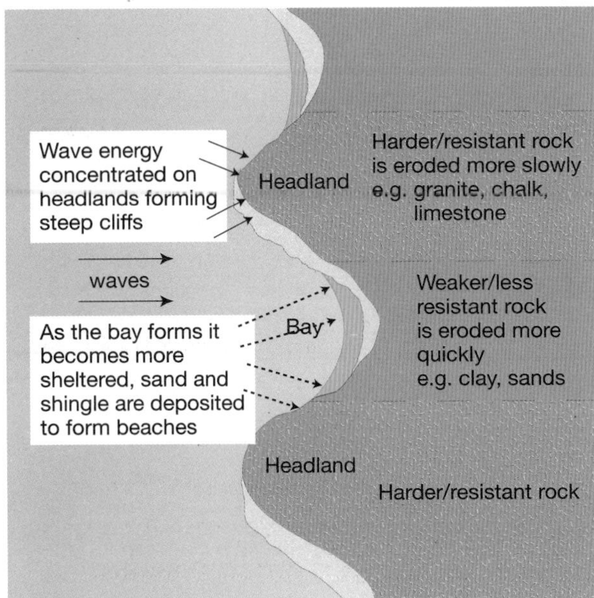

Wave energy concentrated on headlands forming steep cliffs
Headland
Harder/resistant rock is eroded more slowly e.g. granite, chalk, limestone
waves
As the bay forms it becomes more sheltered, sand and shingle are deposited to form beaches
Bay
Weaker/less resistant rock is eroded more quickly e.g. clay, sands
Headland
Harder/resistant rock

Cliffs and wave-cut platforms

Fig 3.4 Stages in the formation of cliffs and wave-cut platforms

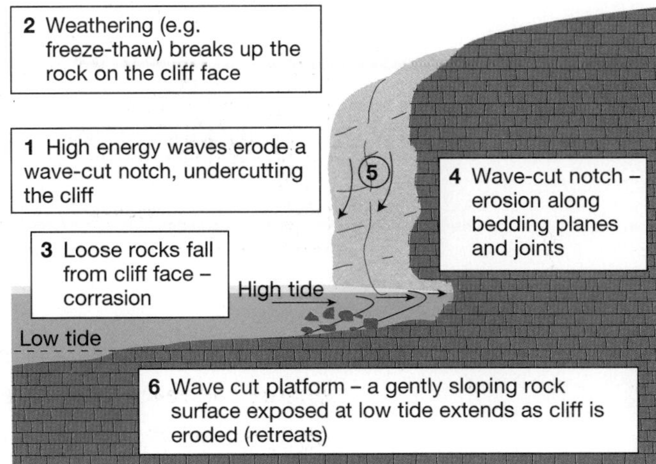

5 Cliffs eventually collapse (mass movement) as they are undercut and retreat
2 Weathering (e.g. freeze-thaw) breaks up the rock on the cliff face
1 High energy waves erode a wave-cut notch, undercutting the cliff
5
4 Wave-cut notch – erosion along bedding planes and joints
3 Loose rocks fall from cliff face – corrasion
High tide
Low tide
6 Wave cut platform – a gently sloping rock surface exposed at low tide extends as cliff is eroded (retreats)

Link to actual examples you have studied.

Caves, arches, stacks and stumps

Fig 3.5 Formation of caves, arches, stacks and stumps, e.g. Fig 3.6 Isle of Purbeck (Old Harry and His Wife) and Fig 3.13 Flamborough Head

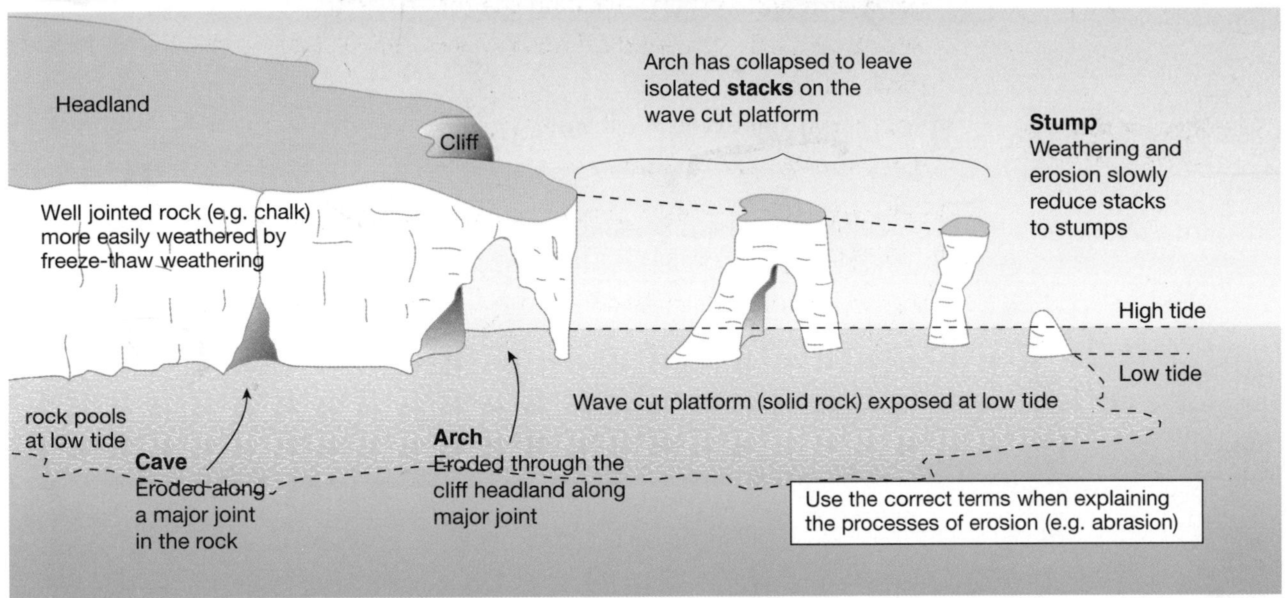

Headland
Cliff
Arch has collapsed to leave isolated **stacks** on the wave cut platform
Stump Weathering and erosion slowly reduce stacks to stumps
Well jointed rock (e.g. chalk) more easily weathered by freeze-thaw weathering
High tide
Low tide
Wave cut platform (solid rock) exposed at low tide
rock pools at low tide
Cave Eroded along a major joint in the rock
Arch Eroded through the cliff headland along major joint
Use the correct terms when explaining the processes of erosion (e.g. abrasion)

Case study: Coastal landforms, geology and processes along the Dorset coast

Fig 3.6 Coastal landforms, geology and processes along the Dorset coast (OS 1:50 000 maps 194 and 195)

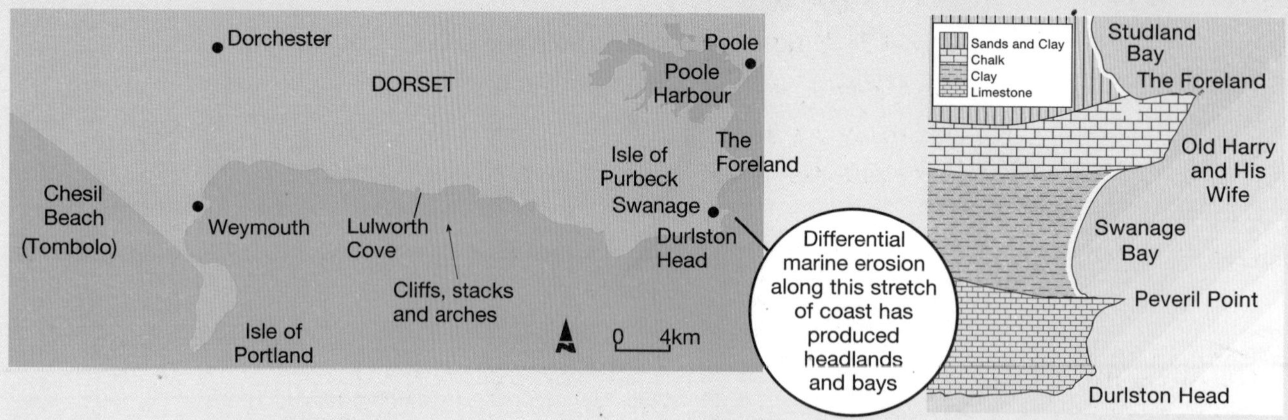

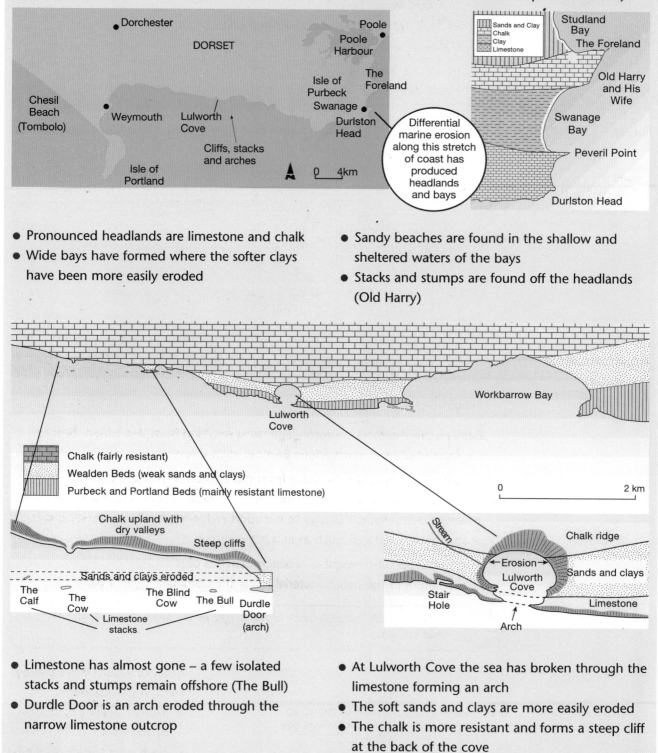

- Pronounced headlands are limestone and chalk
- Wide bays have formed where the softer clays have been more easily eroded
- Sandy beaches are found in the shallow and sheltered waters of the bays
- Stacks and stumps are found off the headlands (Old Harry)

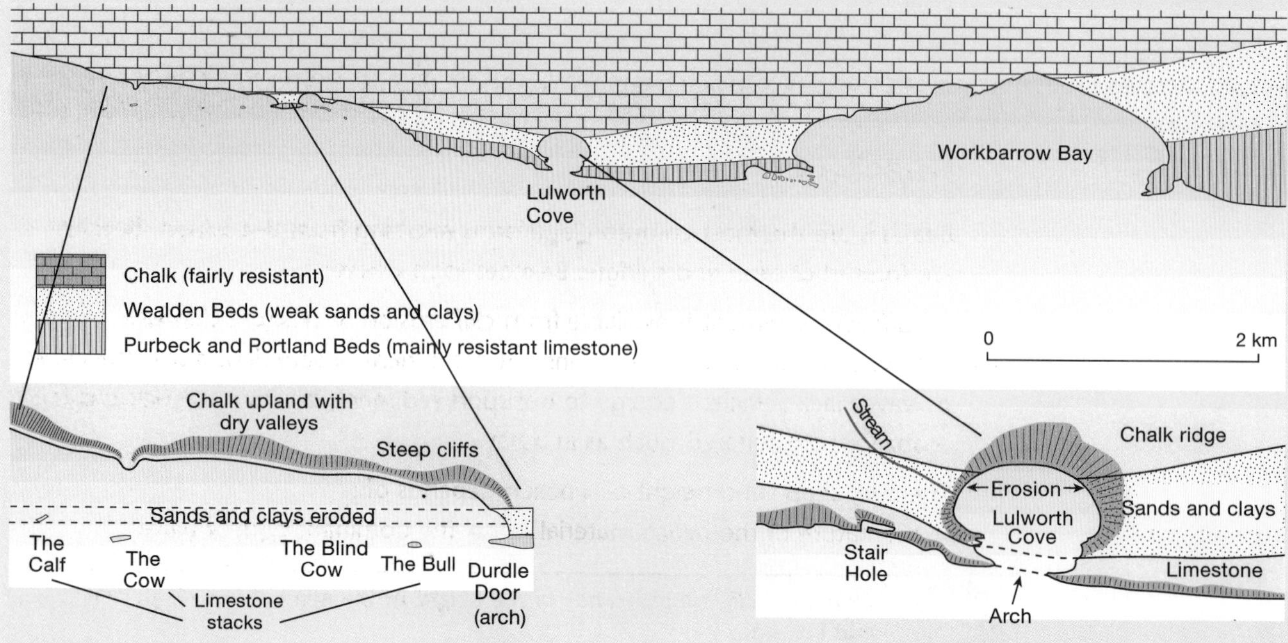

- Limestone has almost gone – a few isolated stacks and stumps remain offshore (The Bull)
- Durdle Door is an arch eroded through the narrow limestone outcrop
- At Lulworth Cove the sea has broken through the limestone forming an arch
- The soft sands and clays are more easily eroded
- The chalk is more resistant and forms a steep cliff at the back of the cove

PROGRESS CHECK

Draw a labelled diagram to show how Workbarrow Bay has been formed.

Processes of transportation

AQA A AQA C

Waves transport:

- rocks and sediments eroded from the cliffs
- sediments carried down to the sea by rivers.

Sediments are moved up and down the beach (swash and backwash). Sediments are moved along the beach by **longshore drift**.

Fig 3.7 Longshore drift (see Figs 3.9–3.11)

Dominant waves come from the direction of maximum fetch and the strongest or prevailing winds.

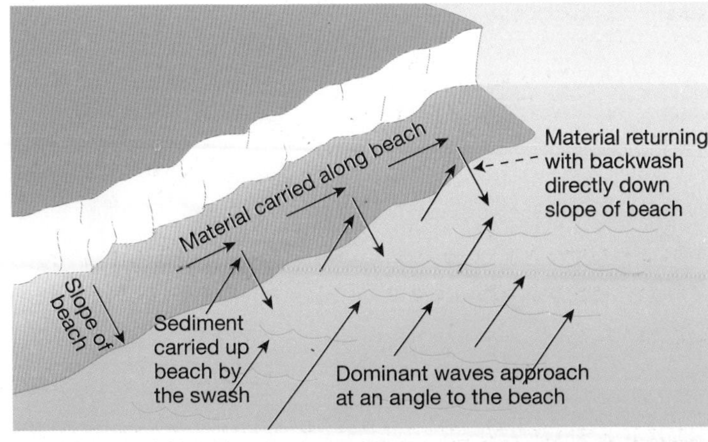

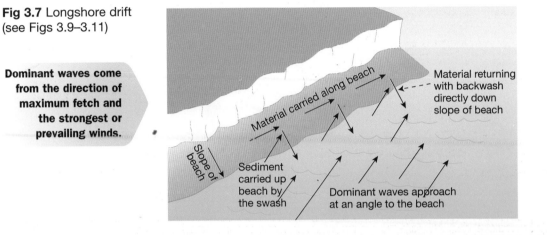

Landforms of deposition

AQA A AQA B

Beaches

Beaches are the most common landforms resulting from deposition. Beaches are formed of sand and shingle. Beaches form where:

- sufficient sediment is available from cliff erosion or offshore deposits
- longshore drift maintains a constant movement of sediment along the coast
- waves lack sufficient energy to transport sediment, because the water is too shallow or sheltered, such as in a bay.

The size, shape and height of a beach depends on:

- the nature of the beach material
- the dominant type of wave

KEY POINT **Tidal range is the difference in the height of the sea between high tide and low tide.**

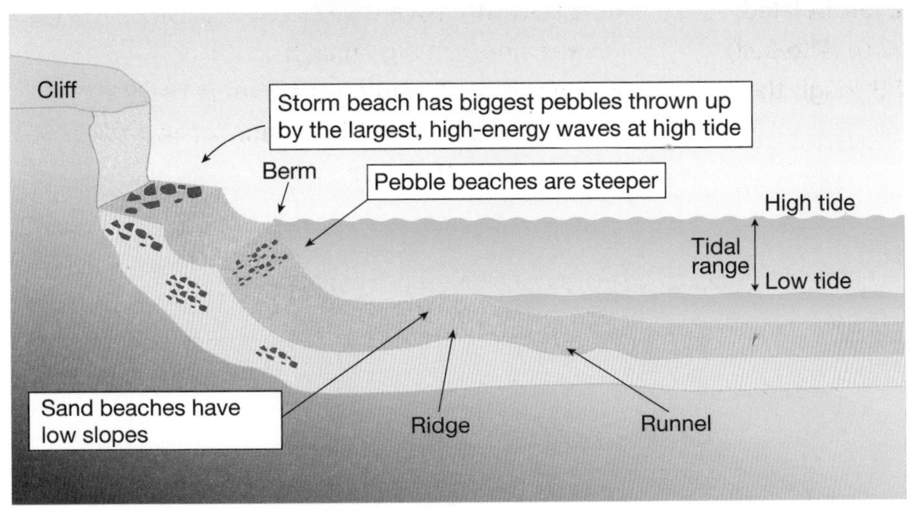

Fig 3.8 Beaches differ in shape, height and size

Beaches absorb wave energy. High, wide beaches:

- protect cliffs from erosion by the waves
- are the best form of coastal defence (e.g. Mappleton Fig 3.13).

Spits

Spits are:

- formed of sand and shingle
- the result of longshore drift
- attached to the mainland at one end
- frequently found where coasts change direction at the mouths of rivers or estuaries.

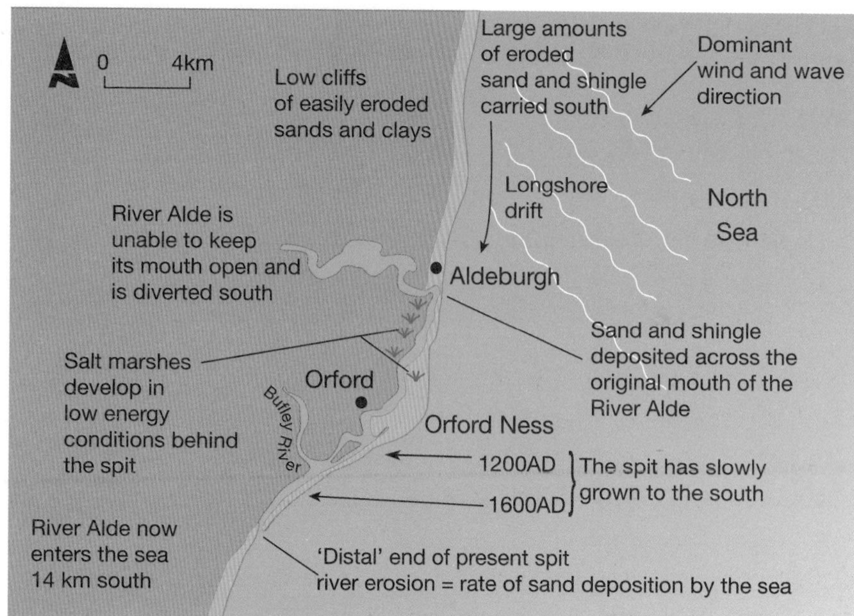

Fig 3.9 Spit formation at Orford Ness, Suffolk (OS 1:50 000 map no. 156 and 169)

Tombolos

Tombolos are similar to spits, but link the mainland to an offshore island as at Chesil Beach in Dorset.

- south-west prevailing winds
- very long fetch
- high energy waves
- shingle beach links coastline to Isle of Portland

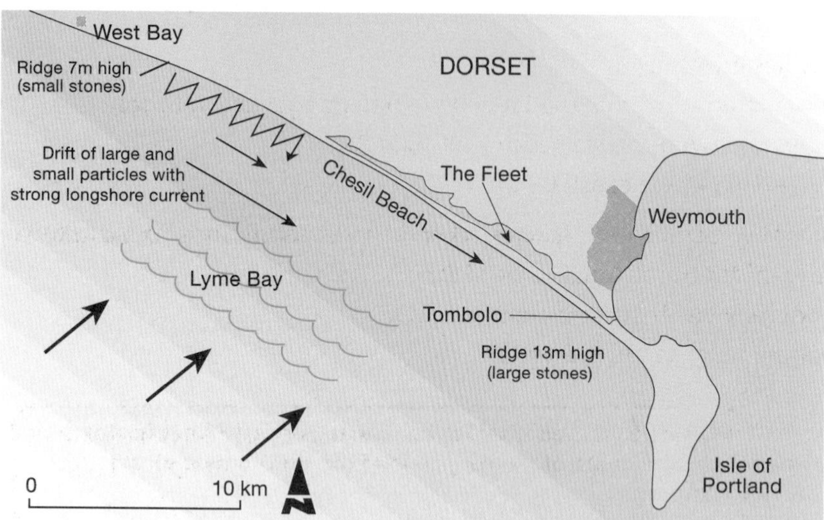

Fig 3.10 Chesil Beach, Dorset (OS 1:50 000 map no. 194)

Bars

Bars are similar to spits, but they form right across a bay, from headland to headland.

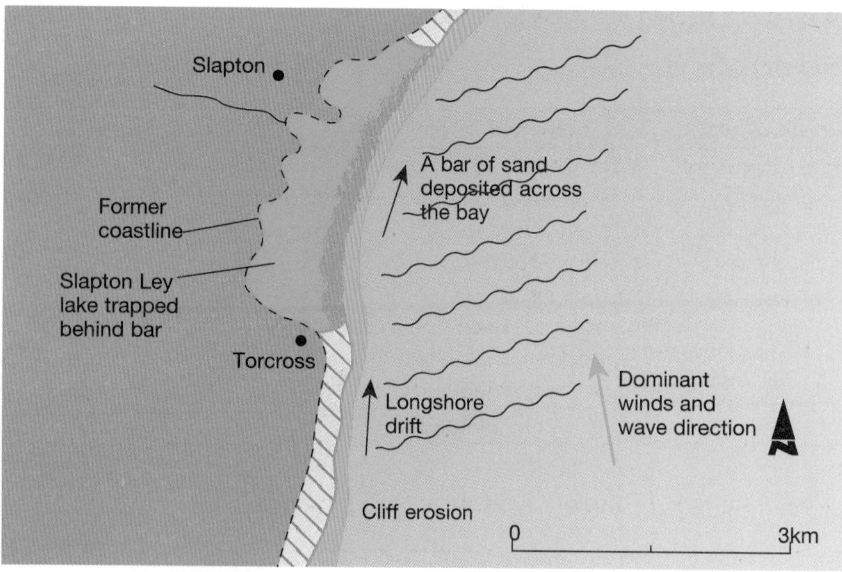

Fig 3.11 Sand bar at Slapton Ley, Devon (OS 1:50 000 map no. 202)

Coastal management

Coasts need to be **managed** to:

- sustain human activities from the threat of coastal erosion
- preserve coasts for heritage and conservation reasons
- preserve coasts from over development.

Managing coastal defences

Coastal defence is necessary in the UK because:

- over 17 million people live within 10km of the coast
- 40% of the manufacturing industry is on or near the coast
- 31% of the coastline has investments in roads, buildings and recreation facilities
- the pressure to develop coastal sites for housing (including holiday and retirement homes) and industry (including tourism and leisure) is likely to increase

Responsibility lies mainly with:

- **The Environment Agency** (national government agency – all coasts)
- **MAFF** (national government agency – all coasts)
- **district councils** (local coasts).

To obtain planning permission, defence schemes must satisfy three basic criteria:

- engineers consider the scheme to be viable
- it is economically viable (cost-benefit analysis)
- it is environmentally sound (sustainable).

Cost-benefit analysis assesses whether the benefits gained from having a sea defence (e.g. the costs of flooding, loss of life etc.) outweigh the costs of construction.

Planning options are shown in Table 3.1 below.

Table 3.1 Planning options

Do nothing	Prevent and discourage	Managed retreat	Build defences
Allow natural processes of erosion to continue, until a new balance has been achieved. Unlikely option where urban areas need defending.	Planning controls prevent further development at vulnerable sites. Mortgages refused and cost of insurance increased.	Defend the coast further inland and allow the present coast to be eroded. Provide compensation and resettlement schemes for farmers and house owners etc. Viable only in areas of low population?	'**Hard**' defences such as sea walls, rip-rap, groynes. '**Soft**' defences such as beach nourishment.

Types of coastal defence are shown in Table 3.2 on page 42.

Use examples from a coast you have studied to describe and explain:

- how defences help reduce erosion
- the advantages and disadvantages for people (e.g. tourists, pleasure boat users and fishermen).

Table 3.2 Types of coastal defence (approximate costs)

Description	Effects	Advantages	Disadvantages
Type: Seawall ('hard' defence)			
Concrete wall of different designs; can include flat 'decks' for walking or car parking. £3000–4000/m	Deflects wave energy back into the next wave.	Capable of withstanding high energy waves. Efficient and effective. Strength and height reassures public fear of flooding.	High cost of building and repair. Reflection can cause turbulence and scouring of the beach. Not very attractive. Access to beach difficult.
Type: Rip-rap (rock armour) ('hard' defence)			
Jumble barrier of large, irregularly shaped rocks or concrete. £3000/m	Absorbs wave energy in gaps and voids between the boulders.	Very efficient. Relatively cheap.	Costs rise if rock imported (e.g. Norway). Rocks must be large enough to remain stable (3 tonnes). Difficult access to beach; dangerous for young children.
Type: Gabions ('hard' defence)			
Strong wire baskets filled with stones and rubble. Often used to protect sand dunes. £100/m	Gaps between stones absorb wave energy.	Trap sand, often become covered with vegetation. Visually less obtrusive.	Not as effective as a sea wall. Shorter life span if exposed.
Type: Groynes ('hard' defence)			
Low walls of timber or concrete built at right angles across the beach. £7000 each	Traps sand moving along the beach by longshore drift. Beach built up between the groynes is able to absorb more wave energy.	Cheaper than sea walls. Maintains beach for tourists. No problems of access	Beaches further along the coast may be starved of a supply of sand and become more vulnerable.
Type: Revetment ('hard' defence)			
Low sloping walls of concrete built at the top of the beach. £2000/m	Reflects and absorbs wave energy.	Beach scouring is less than with a sea wall. Relatively cheap form of 'hard' defence. Less intrusive than a sea wall.	Unsuitable for high energy conditions. Short life span.
Type: Embankment ('hard' defence)			
An earth bank dug from nearby and covered with grass. Often found along banks of tidal estuaries and rivers with a saltmarsh in front.	Prevents water at high tide flooding the low-lying land behind. Only found along coasts where wave energy is low.	Relatively low cost. Cheap maintenance.	Unable to withstand high energy waves. Erodes quickly if water overflows the bank.
Type: Beach nourishment ('soft' defence)			
Beach made higher and wider by feeding sand and shingle brought in by lorries or dredgers. Often used to 'protect' hard defences. £20/cu.m	Beach more able to absorb wave energy, particularly in storm conditions.	Very effective. Relatively cheap. Maintains natural appearance of coast. Preserves a beach for leisure purposes.	Beaches probably need to be re-nourished if storms cause erosion, increasing the cost. Offshore dredging may increase erosion in another location.

Fig 3.12 Flood risk and beach nourishment sites along the east coast

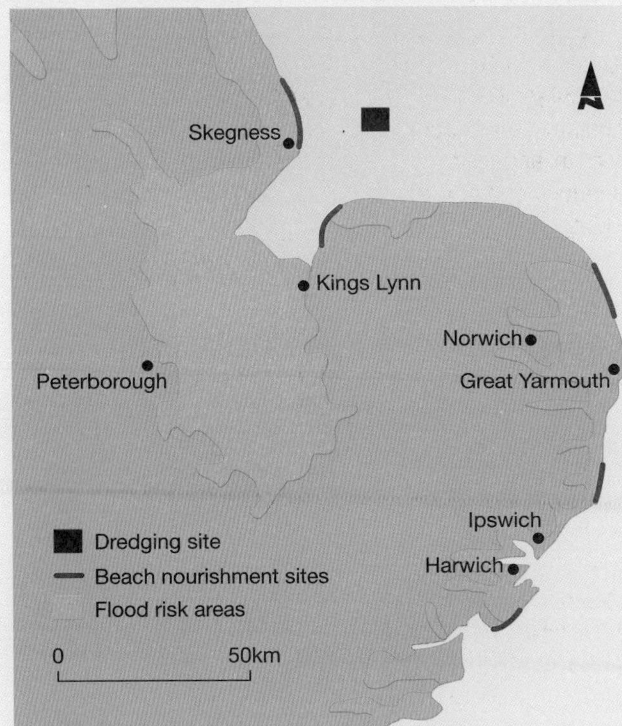

Case study: Holderness and Lincolnshire coasts

The east coasts of the UK south of Flamborough Head (Yorkshire) are vulnerable to erosion and flooding because:

- much of the area is low lying with many parts less than 5m above average sea-level
- cliffs are formed of soft, easily-eroded rocks such as glacial till (boulder clay)
- East Anglia is slowly sinking relative to sea-level
- global warming is likely to cause:
    - a rise in sea-level, particularly threatening low-lying coasts
    - increased storminess in the North Sea and more storm surges.

Fig 3.13 Management problems and solutions along the Holderness and Lincolnshire coasts

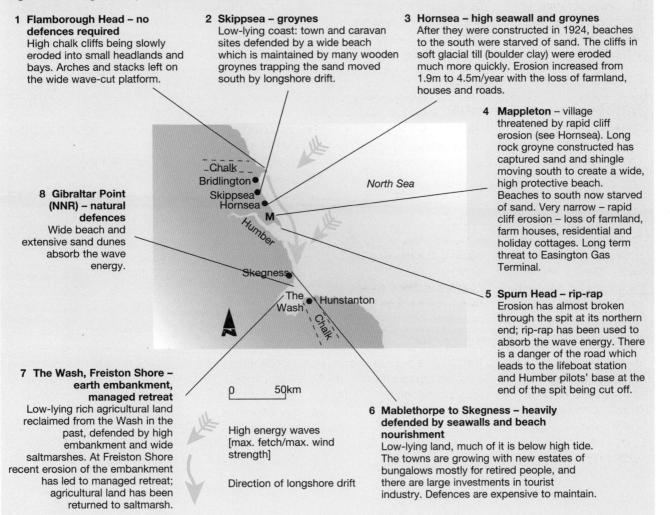

1 Flamborough Head – no defences required
High chalk cliffs being slowly eroded into small headlands and bays. Arches and stacks left on the wide wave-cut platform.

2 Skippsea – groynes
Low-lying coast: town and caravan sites defended by a wide beach which is maintained by many wooden groynes trapping the sand moved south by longshore drift.

3 Hornsea – high seawall and groynes
After they were constructed in 1924, beaches to the south were starved of sand. The cliffs in soft glacial till (boulder clay) were eroded much more quickly. Erosion increased from 1.9m to 4.5m/year with the loss of farmland, houses and roads.

4 Mappleton – village threatened by rapid cliff erosion (see Hornsea). Long rock groyne constructed has captured sand and shingle moving south to create a wide, high protective beach. Beaches to south now starved of sand. Very narrow – rapid cliff erosion – loss of farmland, farm houses, residential and holiday cottages. Long term threat to Easington Gas Terminal.

5 Spurn Head – rip-rap
Erosion has almost broken through the spit at its northern end; rip-rap has been used to absorb the wave energy. There is a danger of the road which leads to the lifeboat station and Humber pilots' base at the end of the spit being cut off.

6 Mablethorpe to Skegness – heavily defended by seawalls and beach nourishment
Low-lying land, much of it is below high tide. The towns are growing with new estates of bungalows mostly for retired people, and there are large investments in tourist industry. Defences are expensive to maintain.

7 The Wash, Freiston Shore – earth embankment, managed retreat
Low-lying rich agricultural land reclaimed from the Wash in the past, defended by high embankment and wide saltmarshes. At Freiston Shore recent erosion of the embankment has led to managed retreat; agricultural land has been returned to saltmarsh.

8 Gibraltar Point (NNR) – natural defences
Wide beach and extensive sand dunes absorb the wave energy.

Chalk
Bridlington
Skippsea
Hornsea
M
Humber
North Sea
Skegness
The Wash
Hunstanton
Chalk
0 50km

High energy waves [max. fetch/max. wind strength]

Direction of longshore drift

Managing coasts for heritage and conservation

Case study: Dorset World Heritage Coast (Jurassic Coast)

Fig 3.14 Conflicting demands on a coastline

Continual demand for more caravan and camping sites
- pressure on footpaths
- disturbance to wildlife

Important fishing grounds sustaining 400 boats
International venue for watersports

17 million tourist visitors per year
- 200,000 educational visitors per year
- overcrowded honeypot sites (Lulworth Cove)
- pressure on car parks, amenities and footpaths

UK's largest onshore oilfield
- infrastructure concealed by trees etc
- expensive underground pipeline

Port and ferry services
Fishing
- danger of marine pollution
- provides employment

Offshore sand and gravel reserves
- removal can change wave patterns and erosion of coast
- conflict with fishing, marine archaeology

AONB (Area of Outstanding Natural Beauty)

Heritage Coast

Protected geological and wildlife sites

World Heritage site

Holiday resort Fishing
Lyme Regis
Axmouth
Golden Cap (Highest cliff in S. Eng)
Seatown
Dorchester
Poole
Studland
NNR
Swanage
The Fleet (wetland site)
Weymouth
Lulworth Cove
Portland shipwreck centre

Research how towns such as Lyme Regis use the 'Jurassic Coast' to attract tourists.

Long distance coastal footpath
- footpath erosion on cliff top can lead to increase in cliff retreat and costly maintenance, e.g. National Trust at Golden Cap

Lyme Bay marine conservation areas
- possible conflicts with:
 - fishing
 - water sports
 - marine archaeology

Coastline of great beauty and rich cultural heritage
- need for interpretation centres, e.g. fossil collecting sites

Important for marine archaeology based on 600 known shipwrecks

Outstanding scientific and educational value (geology and landforms)
- overuse for field trips can increase erosion

Exceptional range of wildlife habitats (heath and wetland) in National Nature Reserve

KEY POINT — 'Honeypot sites' are places that attract very large numbers of visitors especially at weekends and Bank Holidays, e.g. Lulworth Cove, Stratford-upon-Avon.

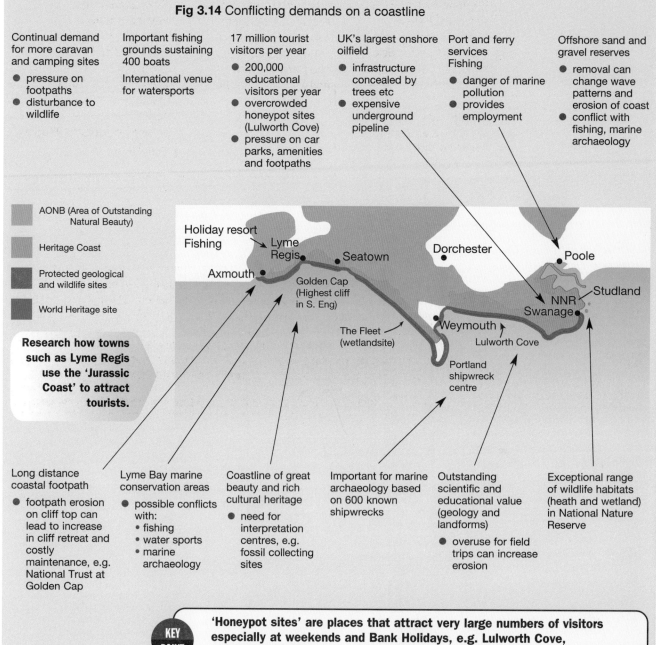

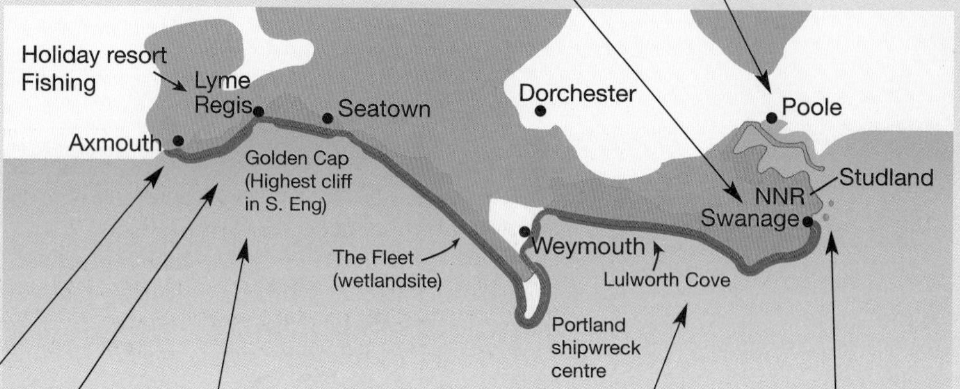

Planning to sustain an environment  **The decision makers**

Heritage Coast Status Area of Outstanding National Beauty (AONB) Sites of Special Scientific Interest (SSSI) National Nature Reserves (NNR) Special Marine Conservation Areas RAMSAR Sites Wetland Sites	Planning regulations to sustain an environment, its natural beauty, and to conserve special areas, sites and habitats of national and international importance	Environment Agency English Nature National government National Trust EU (projects and grants) Dorset County Council Local District Councils MAFF Ministry of Culture, Media and Sport

1. What name is given to the distance the wind has travelled over the sea?
2. Name the process of erosion when wave pressure breaks up rocks.
3. What is the most common landform produced by deposition?
4. Orford Ness is an example of which type of landform?
5. Is beach nourishment a 'hard' or 'soft' type of sea defence?
6. Which area of the UK has large areas at risk from coastal flooding?
7. What is the name given to a place that attracts very large numbers of visitors?

1. Fetch 2. Hydraulic action 3. Beach 4. Spit 5. 'Soft' 6. South east 7. Honeypot

Chapter 4 Glacial landscapes

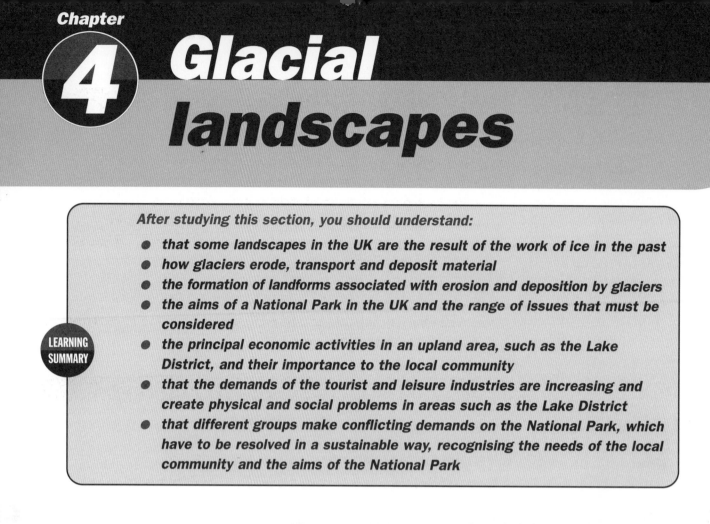

After studying this section, you should understand:

- ***that some landscapes in the UK are the result of the work of ice in the past***
- ***how glaciers erode, transport and deposit material***
- ***the formation of landforms associated with erosion and deposition by glaciers***
- ***the aims of a National Park in the UK and the range of issues that must be considered***
- ***the principal economic activities in an upland area, such as the Lake District, and their importance to the local community***
- ***that the demands of the tourist and leisure industries are increasing and create physical and social problems in areas such as the Lake District***
- ***that different groups make conflicting demands on the National Park, which have to be resolved in a sustainable way, recognising the needs of the local community and the aims of the National Park***

Glacial landscapes

The spectacular landscapes found in upland areas of the UK such as North Wales, the Cairngorms and the Lake District, are mainly the result of the work of ice in the past. The 'ice age' consisted of many cold periods (**glacials**) separated by periods of warmer temperatures (**interglacials**).

- In a long cold period 30 000 years ago, **icesheets** extended over much of the UK, as far south as a line drawn from Bristol to London.
- During the last glacial period, which ended 10 000 years ago, only the valleys in upland areas were filled with **glaciers**, which flowed down from the highest mountains.

Icesheets flow out of polar and upland areas to cover the whole landscape. Glaciers occupy valleys in upland areas, leaving the higher mountains exposed.

Glacier ice is formed from the accumulation of snow in hollows. The weight of snow compacts the lower layers into névé (ice with many air bubbles). Further compaction and freezing over many years squeezes out the air and forms dense glacier ice.

Movement of glaciers

- As ice accumulates and thickens it becomes heavier.
- The weight of the ice causes it to slide downhill under gravity.
- Movement creates friction with the rocks beneath the ice, which melts a thin layer of ice at the base of the glacier.
- The **meltwater** released acts as a lubricant helping the glacier move faster.

Glacial erosion

AQA A AQA B AQA C

In cold glacial areas, freeze-thaw weathering is very active in breaking up rocks into loose angular fragments. There are two dominant processes of erosion:

- **Abrasion**
  - the ice scoops up loosened rocks as it moves forward
  - loose fragments fall onto the glacier from rocky slopes above
  - the moving ice uses this material to erode and scratch away the side and floor of the valley
- **Plucking**
  - ice freezes to the rocks beneath it
  - when the ice moves it pulls the rock apart along points of weakness (e.g. joints and bedding planes)

Use these terms when explaining the landforms produced by erosion.

Landforms produced by glacial erosion

AQA A AQA B AQA C

Corries (cirques or cwms) which are linked with:

- arêtes
- pyramidal peaks.

Glacial troughs (U-shaped valleys) which are linked with:

- hanging valleys
- truncated spurs
- ribbon lakes

Corries

You should be able to recognise these landforms on an OS map (e.g. OS 1:50 000 map no. 124)

- deep, semi-circular hollows with very steep, precipitous sides or headwalls
- eroded from small hollows high up on the mountain sides, the hollows allow snow to accumulate, forming névé and eventually dense glacier ice
- found in many upland areas such as North Wales, e.g. Cadair Idris.

Fig 4.1 Formation of a corrie

During a glacial period

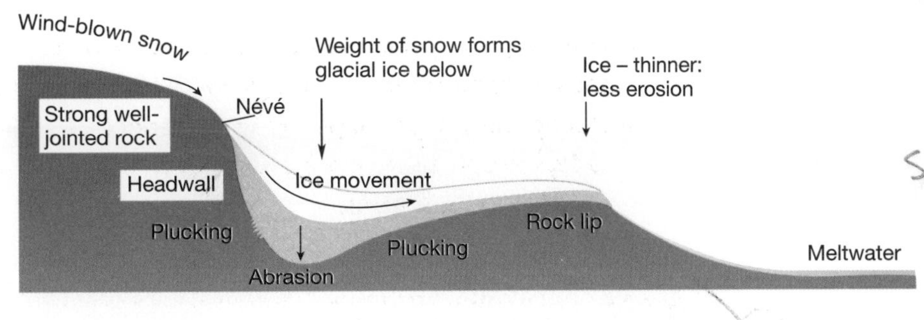

Today

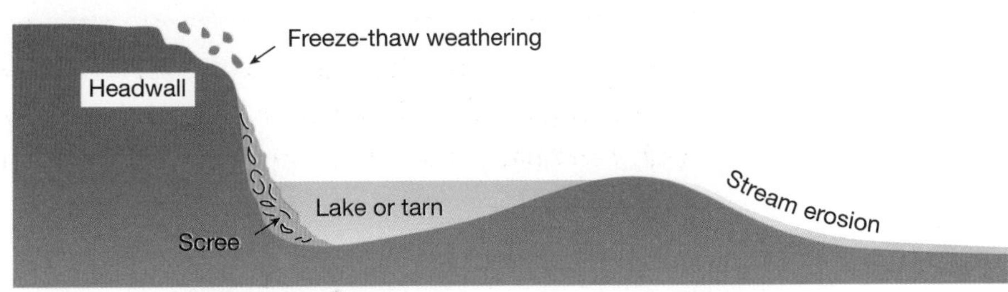

Arêtes are sharp, high ridges of rock where the headwalls of two corries are back-to-back. (Fig 4.3a and 4.3b)

Pyramidal peaks are sharp, angular mountain summits, formed by the erosion of several corries, e.g. Cadair Idris. (Fig 4.3a and 4.3b)

Glacial troughs or U-shaped valleys

- A valley glacier deepens, widens and straightens former river valleys.
- The V-shaped valley is eroded into a U-shape.
- The interlocking spurs are eroded and left as **truncated spurs** forming the steep valley sides.
- Tributary valleys not deepened as much as the main valley are left as **hanging valleys**, high up on the valley sides, marked today by waterfalls or steep gorges.
- Softer, less resistant rocks along the main valley floor were excavated more deeply and are now filled with water and called **ribbon lakes**.

Fig 4.2 Formation of a glacial trough

(a) Before glaciation

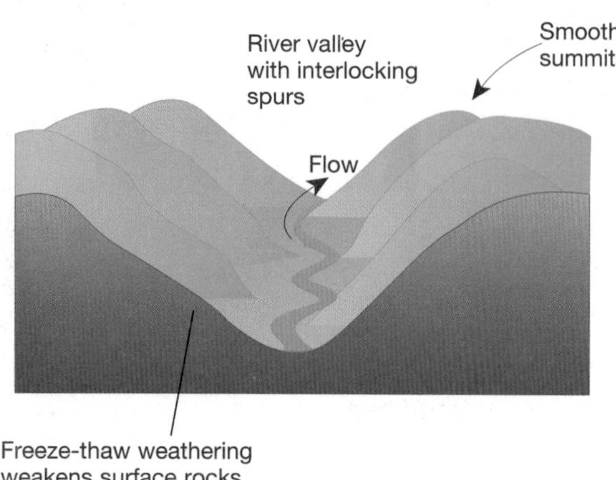

(b) During glaciation

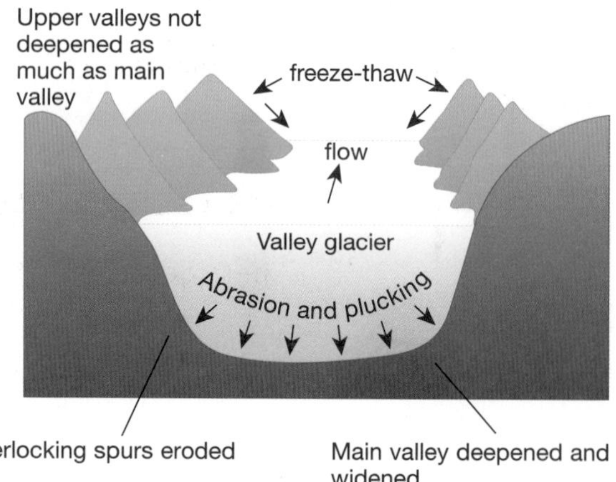

(c) Today

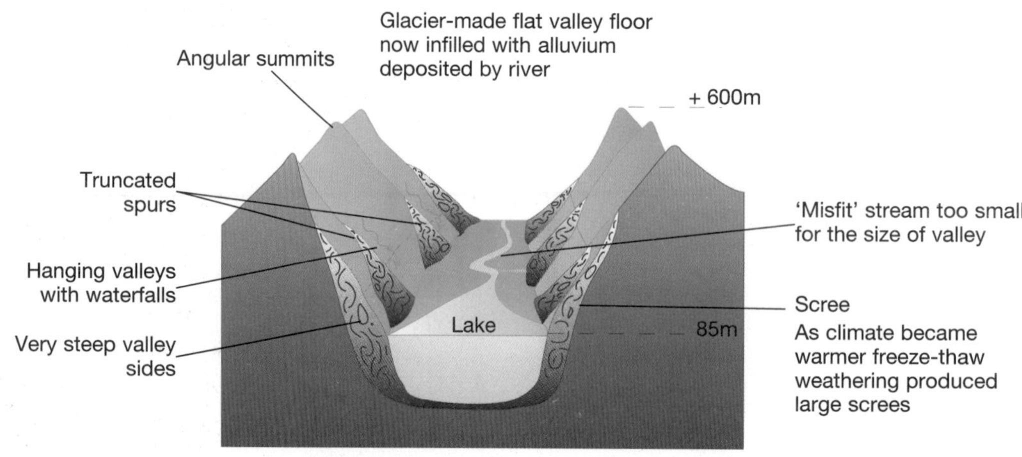

Case study of Cadair Idris, Wales

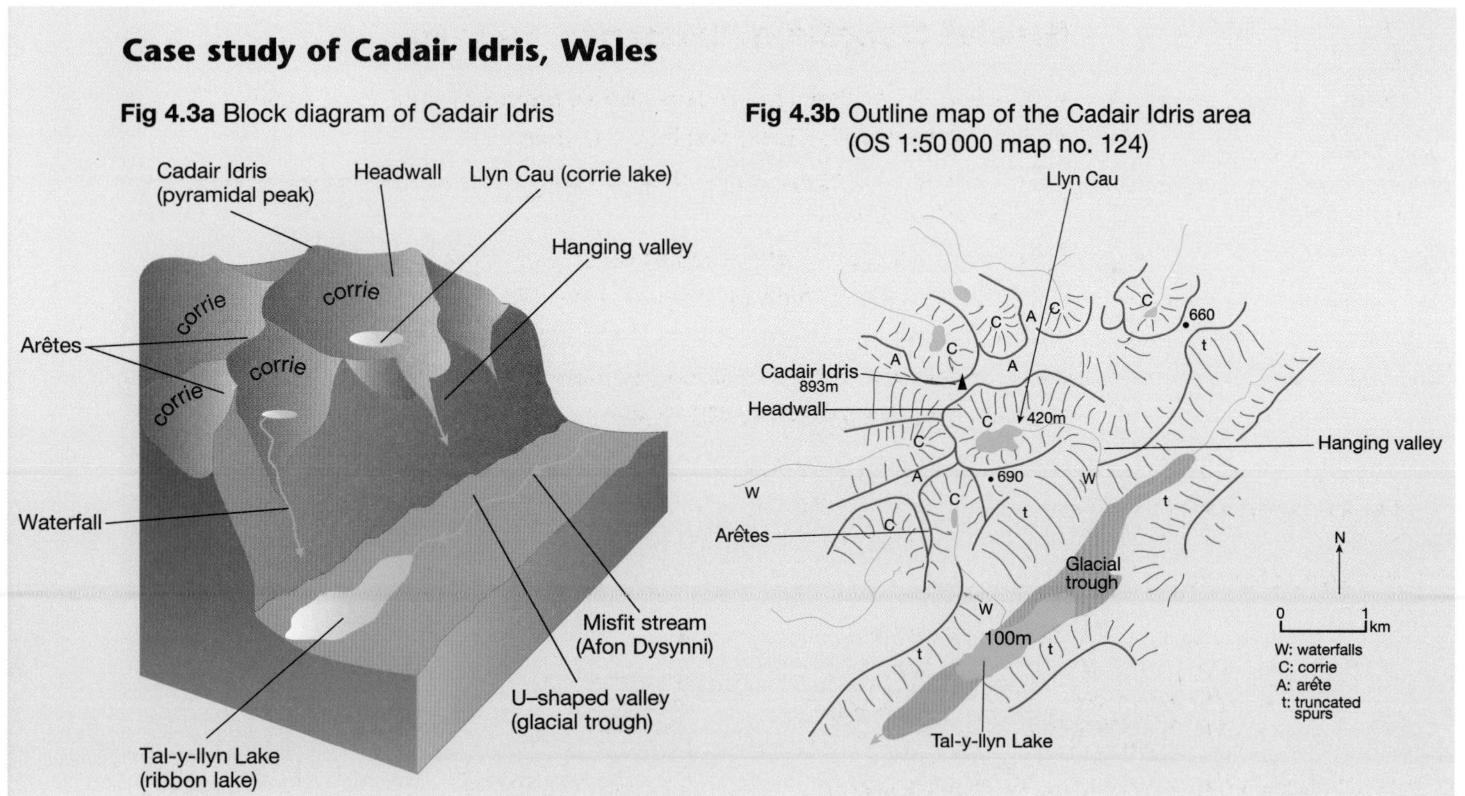

Fig 4.3a Block diagram of Cadair Idris

Fig 4.3b Outline map of the Cadair Idris area (OS 1:50 000 map no. 124)

Transportation by ice

The large amount of rock carried by a glacier is known as its **load**. This consists of:

- angular fragments of rock
- large boulders to fine clay (unsorted).

Fig 4.4 Formation of lateral and medial moraines

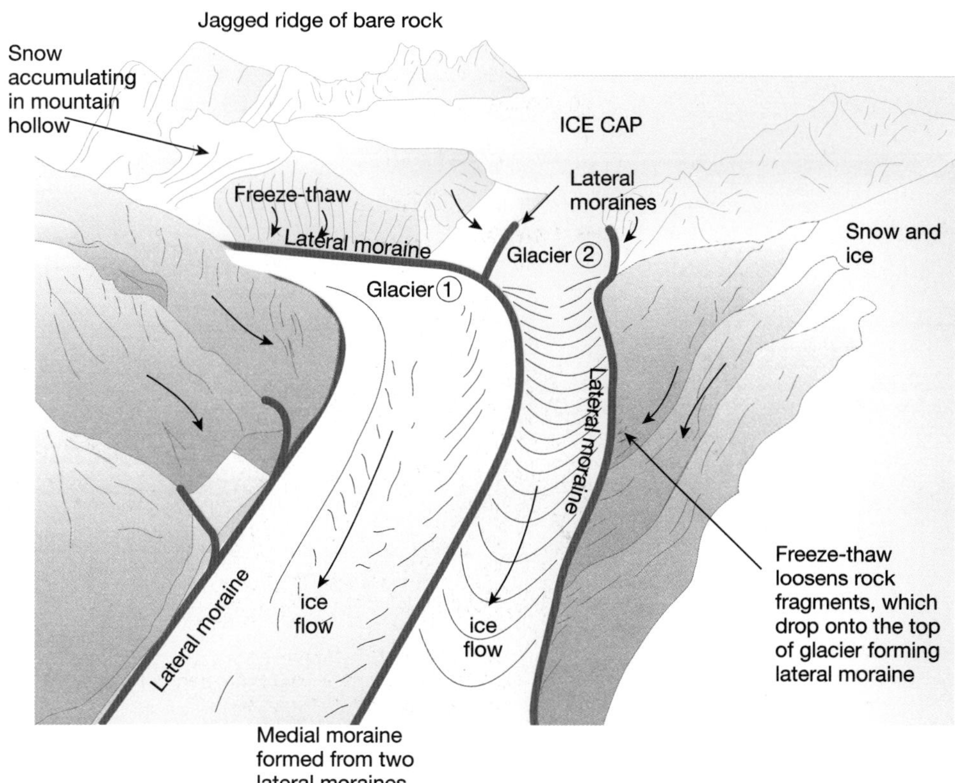

Landforms produced by glacial deposition

AQA A AQA B AQA C

The end of the last ice age was marked by:
- gradual increase of global temperatures and lower snowfall
- glaciers melted faster than the rate at which they moved forward
- glaciers became smaller, retreated and slowly disappeared from the UK.

Glacial till is the more accurate and scientific term.

The material carried by the ice was left behind as **moraine**, deposited on the rocks and landscape below the ice. These deposits are known as **glacial till** or **boulder clay**. Glacial till is spread like a blanket over much of lowland UK, particularly in eastern areas such as Holderness, Lincolnshire and East Anglia.

Landforms formed of glacial till include:

- **terminal moraines** – low ridges of till, marking the furthest position reached by the ice front
- **lateral moraines** – low ridges of till, along the side of the valley floor, formed from rocks falling from the mountain sides above the glacier
- **drumlins** – streamlined shapes of till, formed under the ice, face up-valley. Origin still unclear. Often found in groups such as in the Eden Valley, Cumbria (OS 1:50 000 map no. 91).
- **erratics** – larger rocks transported by the ice and left in an area of very different rocks, e.g. rocks from Norway are found in East Anglia.

Fig 4.5 Landforms of glacial deposition

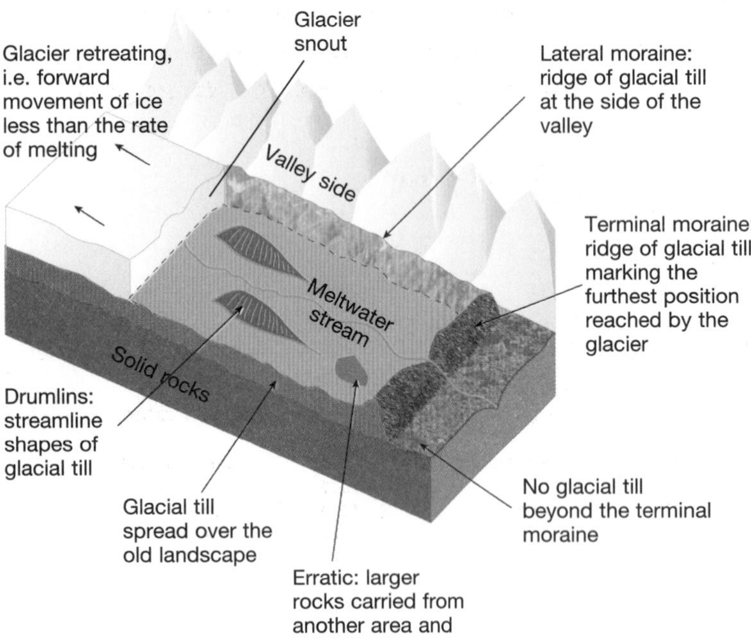

Human activities in an upland glaciated area

AQA A AQA B AQA C

KEY POINT

A National Park is a local government body administered mainly by local councillors and managed by an appointed National Park Officer and staff. (See also pages 187–8.)

Case study: the Lake District National Park

The Lake District National Park (see Fig. 4.8) is run by the Lake District National Park Authority (LDNPA) with the following aims:
- to conserve and enhance the natural beauty of the area
- to encourage and promote opportunities for people to enjoy leisure activities in the area and its natural beauty
- to manage and improve the quality of life of people living there (e.g. employment, housing, services).

Locate the Lake District in your atlas and identify some of these places.

Characteristics of the Lake District National Park:

- High mountain areas, such as Coniston Fell (800m), are separated from each other by deep, steep-sided, U-shaped valleys.
- Many valleys are partly filled by long, deep, ribbon lakes such as Coniston Water, Windermere and Ullswater.
- Much of the scenery and landforms are the result of the work of ice and glaciers in the past.
- Very high rainfall (Coniston has 2200mm/year compared with London 600mm/year) and impermeable rocks give rise to many tarns, mountain streams, waterfalls and lakes.
- The climate is characterised by cool summers (15°C compared with London, 22°C in July) and mild winters (5°C compared with London, 7°C in January).
- The most important land-use is pastoral farming, raising sheep and cattle.
- Large upland areas and steep valley sides are planted with forests for commercial use; 75 000 tonnes of timber are cut and sold each year.
- Communications are difficult; main roads mostly follow valleys.
- Tourism is very important and increasing (1998: 12 million visitors/year)

Fig 4.6 Sketch cross-section of the Coniston Valley (based on OS 1:50 000 map no. 97)

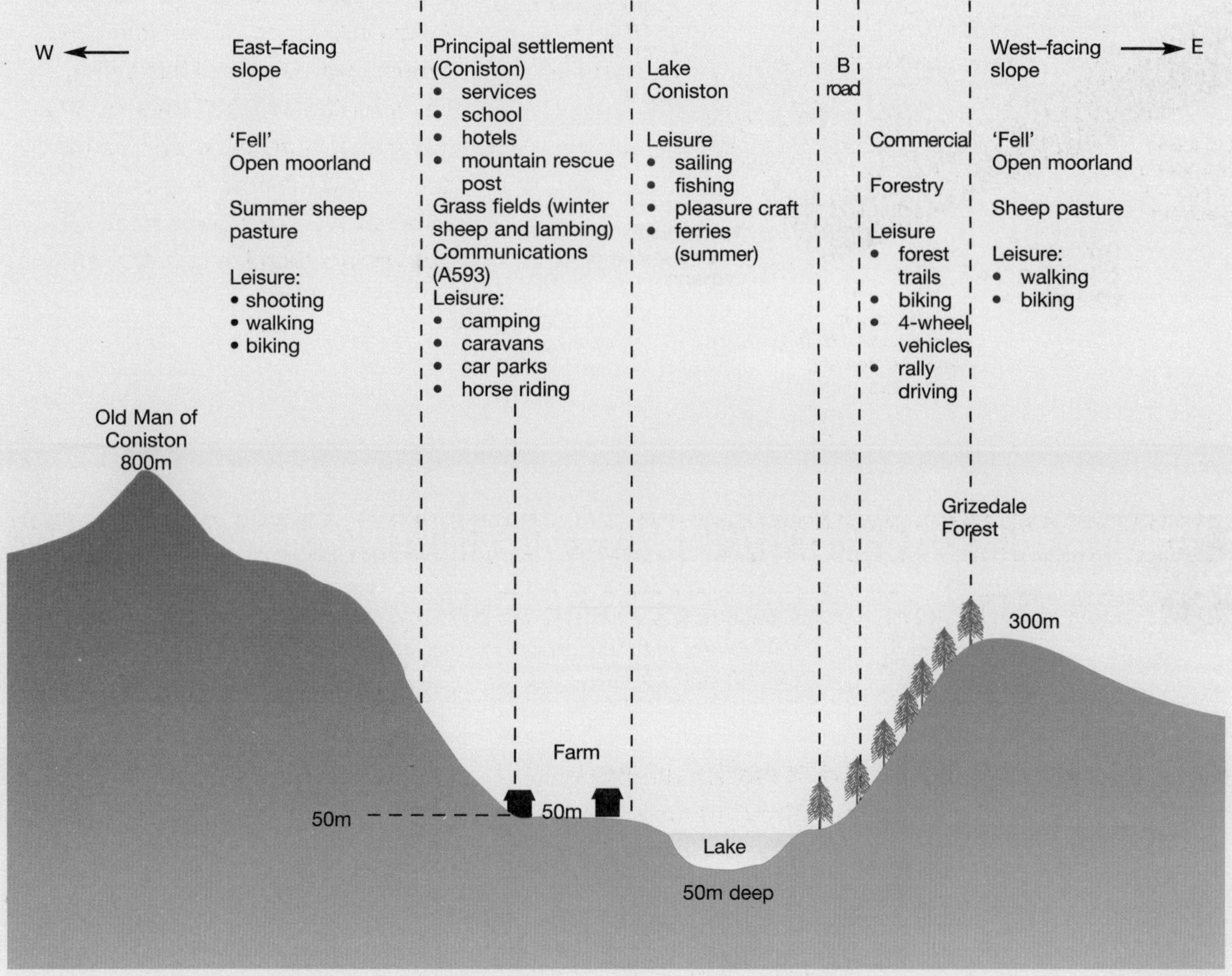

Case study: Hill farming in the Lake District (see also page 150)

Hill farming produces lambs and calves. These are sold on to lowland farms in other areas of the country for fattening (meat) and breeding.

Fig 4.7 The farm system

Inputs →	Processes →	Outputs
Physical: • High rainfall • Low sunshine hours (cloudy) • High fells (upland areas) • Thin, poor soils on the fells • Steep valley sides • Limited, wet, flat land Human: • Family labour • Skilled knowledge Economic: • Low investment (poor profits, uncertain future) • High cost of supplementary feed • Distance from markets and difficult access – high fuel costs • Government policies: High subsidies (CAP) Price support • High cost of fertilisers • Veterinary fees	• Lowland grass management sowing mowing • Silage making (winter feed from grass) • Sheep rearing (lambing) • Cattle rearing (calving) • Fertilising • Bracken control on fells • Maintenance (buildings, stone walls)	• Lambs • Calves

Problems with farming

- Hill farmers have found it difficult to make a liveable profit.
- Many farms have been sold.
- Support from subsidies has led to:
  - over-production, over-stocking and soil erosion
  - quotas on sheep and cattle rearing
- There is a decline in the demand and supply of labour (young people do not want to be farmers).

Possible solutions

- Sell the land to housing or tourism developers (if in a suitable site).
- Seek additional part-time employment, e.g. forestry, National Park wardens.
- Diversify into other areas:
  - farmhouse bed and breakfast
  - conversion to camping barns (planning permission, investment)
  - pony trekking (investment, expertise)
  - quadbike trekking (investment, expertise)
- Seek support to farm the land in a more environment-friendly and sustainable way:
  - Environmentally Sensitive Areas (ESAs) with subsidies to compensate for lower stocking levels and use of less fertiliser
  - management to preserve wildlife, stone buildings and public access
  - Countryside Stewardship – subsidies for conservation and preservation of stone buildings etc.
  - Woodland Grant Scheme – grants to plant and manage woodlands of native species

If you have visited a hill farm on a field trip, revise your notes and use as a case study.

Tourism in the Lake District

AQA A AQA B AQA C

Tourism is a very important part of the local economy. Over 12 million people visit the Lake District each year (compared with a local population of 42 000). The number of visitors is increasing.

Visitors are attracted to the Lake District to:

- enjoy the scenery and landscape
- enjoy the peaceful environment and fresh air
- escape crowds!
- enjoy a variety of leisure pursuits, e.g. walking, water sports, rock climbing, mountain biking
- enjoy the attractive villages and shopping, e.g. craft shops, leisure wear shops

Fifty per cent of visitors come from:

- regions surrounding the Lake District, e.g. Newcastle, Manchester, Leeds
- areas made more easily accessible by the motorway network, e.g. M6 – Birmingham, M74 – Glasgow.

Ninety per cent of visitors travel by car – traffic has increased by over 30% in 20 years.

Visitors come for day visits or stay for weekends and longer holidays:

- about 30% stay in hotels and B&B accommodation
- about 40% rent cottages or bring caravans or tents.

In recent years some controversial proposals include:

- hotel extensions and leisure complexes on lake edges
- timeshare complexes
- marinas
- cable cars (proposal for Helvellyn)
- ski-lifts

Fig 4.8 Ten million people live within a three-hour drive of the Lake District

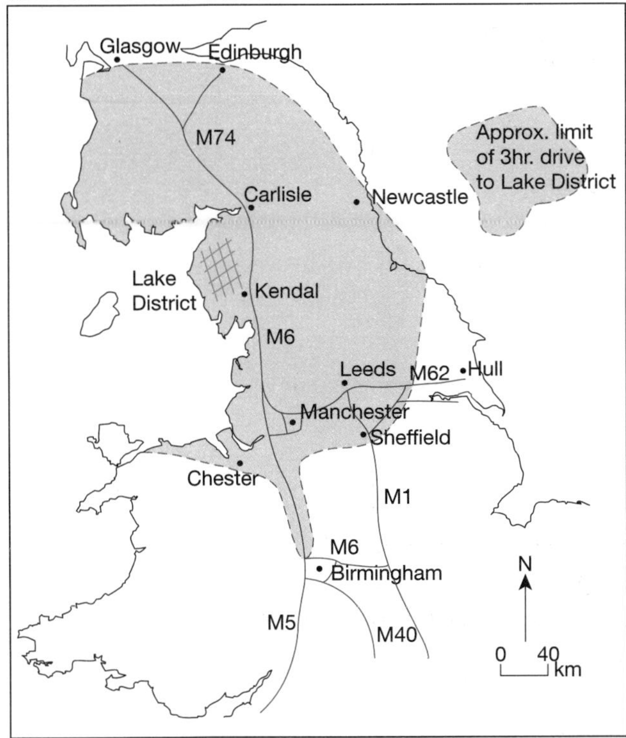

Fig 4.9 Pie graphs comparing employment in the Lake District with the UK as a whole (1991)

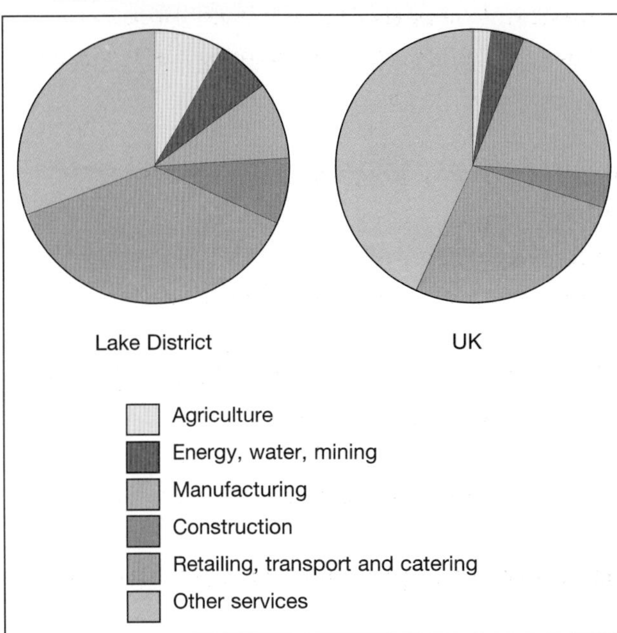

The impact of tourism

Local employment

- In some areas, 50% of the workforce is employed in the tourist industry, e.g. in Windermere: hotels, B&B, shops, cafes, leisure activities (boating).
- Many more jobs rely partly on tourism for business, e.g. builders, bakeries.
- Local tourism employment has helped to compensate for the decline of employment in agriculture and forestry.

Study the pie graphs (Fig 4.9) and make comparisons under the headings:
- **primary industry**
- **secondary industry**
- **service industry.**

Challenges and problems:
- relatively low wages
- **seasonal employment** (up to 50% of jobs)
- **casual labour** often employed from outside Lake District, e.g. students
- restricts development of skills and career
- many young people leave for employment in urban areas, e.g. Manchester.

Solutions:
- policy to recruit staff from local communities
- use local producers and services, e.g. builders
- extend tourist season – guided tours for retired people, discount weekends, theme weekends independent of weather
- encourage local craft industries to develop national markets, e.g. pottery
- encourage **'footloose'** industries (see page 164).

Local services

Questions often ask for the benefits and problems faced by a local community as a result of tourism.

- Employment keeps people in the Lake District, which maintains local services such as schools, shops and leisure centres.
- Loss of 'local' shops to more profitable 'tourist' shops, e.g. gift shops, cafes, leisure wear (Grasmere and Ambleside) – provides employment (seasonal?) but reduces range of 'local' services (increases need to go to larger centres).

Housing

- Housing is in short supply. New housing is restricted to 'infilling' of existing settlements. There is a demand for **second homes**, often small cottages, and people are looking to invest in holiday letting. There is also a demand from outsiders for retirement homes.
- House prices have increased a great deal.
- Local people, particularly young people, cannot afford to buy so move out **(depopulation)** with a consequent threat to local services.

Solutions to housing problems:
- planning permission for new housing only granted if large proportion are sold to local people
- **housing associations** build specifically for young couples.

Traffic and car parking

- The narrow, winding roads within the Lake District are heavily congested in summer, particularly at weekends.
- Car parking in towns, beauty spots and popular walking spots, e.g. Helvellyn, are at saturation point.
- Traffic and parking problems lead to inconvenience to locals, visitor frustration, illegal and dangerous parking.

Solutions to traffic and parking problems:
- restrict number of cars entering the National Park
- enforce more **'park and ride'** schemes into popular areas where there is a single access, e.g. Haweswater
- build speed ramps in villages and restrict coaches, e.g. Grasmere.

Pollution

- There is pollution from cars and coaches particularly when the roads are congested.
- High visitor numbers leads to increased litter.
- There is an increase in sewage effluent in the summer.

Solutions to pollution problems:

- increase recycling in waste management
- educate visitors ('**take litter home**').

Physical environment

Fig 4.10 Stages in footpath erosion

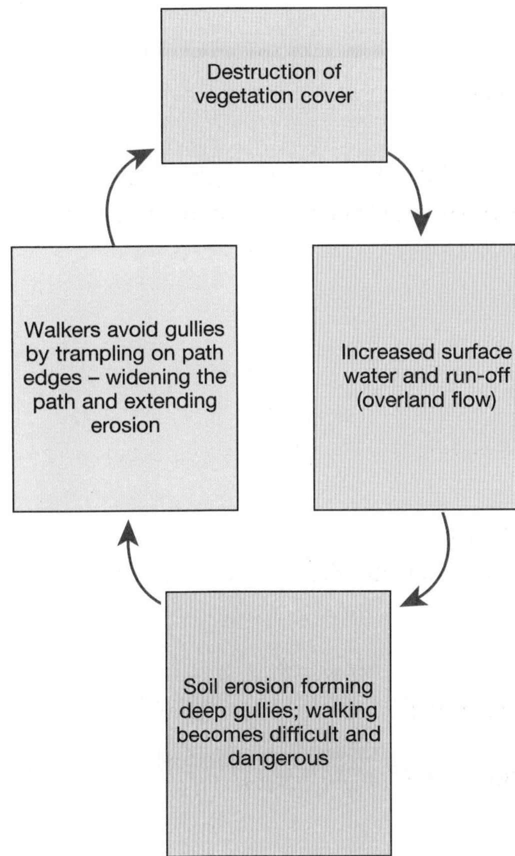
Destruction of vegetation cover
Increased surface water and run-off (overland flow)
Soil erosion forming deep gullies; walking becomes difficult and dangerous
Walkers avoid gullies by trampling on path edges – widening the path and extending erosion

Footpath erosion

- Many access paths to popular mountain tops are so badly eroded that the scars spoil the scenery and the view from below!
- The amount of footpath erosion is influenced by:
  - human factors
    - number of walkers
    - proximity to car parks
    - winter use
  - physical factors
    - amount of rain
    - vegetation cover
    - angle of slope
    - soil compaction
- Repairs and maintenance are expensive in high, remote areas, e.g. £150/metre.

Solutions to problems of footpath erosion:

- obtain grants to carry out repairs to:
  - sympathetically restore the landscape, using local stone
  - provide a safe, hard-wearing path, e.g. paving slabs/duck boards – high cost on remote upland routes (need for helicopters to deliver heavy materials).
- recognise sites liable to erosion in the future, and manage them carefully, e.g. diverting paths so they recover, fencing paths, improving drainage
- educate walkers.

There are also problems and conflicts arising from rapidly-growing leisure activities such as mountain biking and four-wheel off-roaders (Table 4.1).

Table 4.1 Issues of mountain biking and off-road leisure vehicles (4-wheel drive)

Mountain biking		Off-road leisure vehicles	
Problems	**Solutions**	**Problems**	**Solutions**
Rapidly growing activity – great demand	Establish more routes at low levels	Increasing ownership but limited knowledge of driving techniques	Establish a code of practice More training schools
Increases erosion of tracks and paths	Separate bikers from walkers	Many tracks made impassable	Classify routes based partly on difficulty
Conflict between bikers and walkers/farmers	Better signposting and education (cycle users code)	Extra noise pollution	Restrict access

Fig 4.11 Management issues around Lake Bassenthwaite [based on OS 1:50 000 map no. 90]

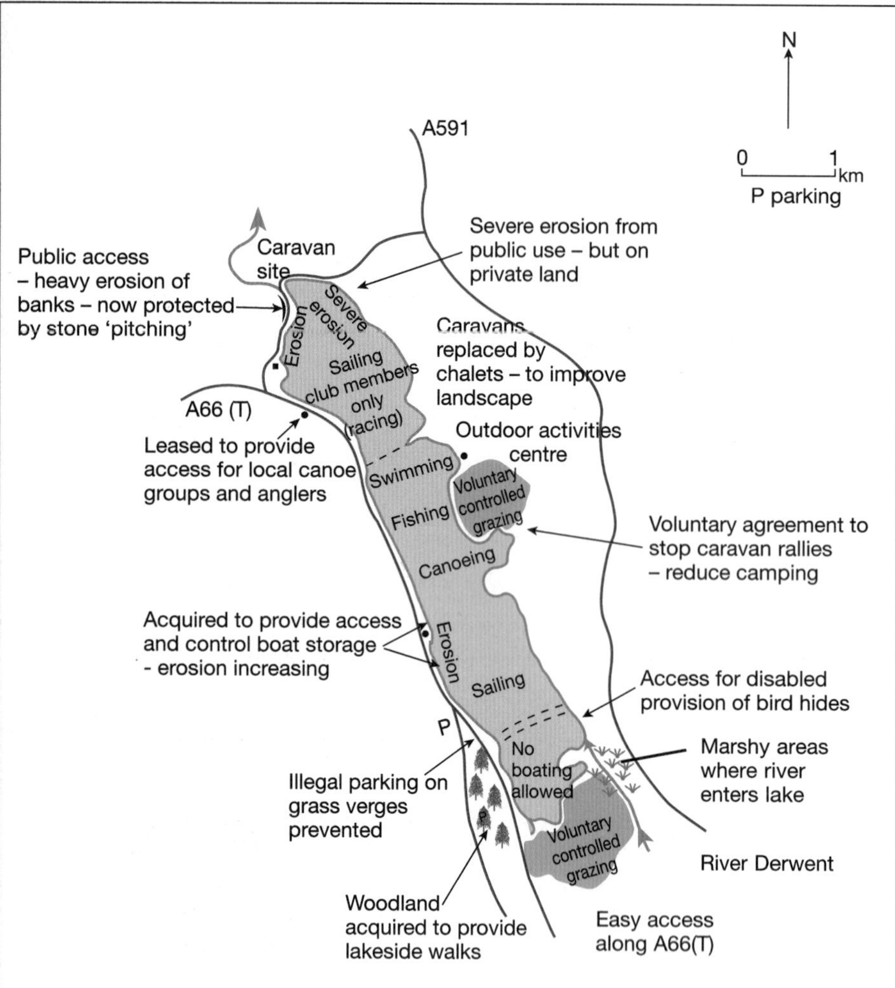

Use of lakes for leisure

- The increased use of lakes for water skiing, power boating, jet skiing has increased erosion of the lake edges.
- Conflicts have arisen between power boats, sailors, anglers, swimmers.
- There are problems of pollution and noise.

Solutions to problems associated with using the lakes for leisure:

- restrict use – in 2005 a 10 mph speed limit was introduced on Lake Windermere (Research the conflicts this causes.)
- segregate uses, e.g. Lake Bassenthwaite (see Fig 4.11).

Sustainable tourism

Partnership and co-operation between the LDNPA, the National Trust, the hotel and leisure industry, and conservation groups are encouraging more sustainable tourism. This may be achieved by:

- encouraging the conservation of the landscape (restricting the number of cars?)
- raising funds to conserve and restore the landscape (entry fees to the Lake District?)
- raising awareness of visitors (more visitor centres, wardens?)
- showing that tourism and conservation must support each other if they are to be successful (too many people and cars destroy the features that attract the tourists)
- ensuring that tourism and conservation benefit the local community (employment, homes, services).

Consider how different groups using the Lake District (e.g. local farmers, hotel owners, residents, tourists) might view the disadvantages/ advantages of each policy.

PROGRESS CHECK

1. Which weathering process is very active in glaciated areas?
2. What is the name given to sharp, high ridges of rock where the headwalls of two corries are back-to-back?
3. What are drumlins made of?
4. What is the name given to the ridge of morainic material marking the furthest extent of a glacier?
5. Name one of the outputs of a hill farm in the Lake District.
6. What industry offers the main employment for people living in the Lake District?

1. Freeze–thaw 2. Arêtes 3. Glacial till or boulder clay 4. Terminal moraine 5. Lambs or calves 6. Tourism

Chapter 5

Weather and climate

LEARNING SUMMARY

After studying this section, you should understand:

- ***the differences and relationships between weather and climate***
- ***that solar radiation is the main source of energy in the atmosphere and that variations in solar energy over time (seasons) and space (latitude) are important in explaining weather and climatic patterns***
- ***how atmospheric conditions can be measured and the processes explained***
- ***how depressions and anticyclones affect the weather of the UK***
- ***the factors affecting climate***
- ***how to describe and compare climates in different parts of the world***
- ***the causes and effects of tropical storms, drought and desertification***
- ***that atmospheric conditions may be changing as a result of human activities, which will have important social, economic and political consequences***

Weather

AQA C

Weather describes the atmospheric conditions at a particular place and time together with the changes taking place over the short term (hour-by-hour or day-by-day).

Recording the weather

Data on the weather is obtained from:

- daily readings of instruments at a weather station
- automatic instruments in remote locations, e.g. ocean buoys
- ships and aeroplanes
- satellites and radar images.

Meteorologists are responsible for recording and analysing the weather. Data is recorded at least once every 24 hours. The information is plotted on maps or synoptic charts, then analysed to obtain a weather forecast.

A range of data is recorded

Temperature

- Air temperature is recorded in the shade (in a Stevenson's Screen).
- It is recorded using a **maximum/minimum thermometer** (highest and lowest temperatures in the last 24 hours).
- It is recorded in degrees **centigrade** (Celsius) or **Fahrenheit**.

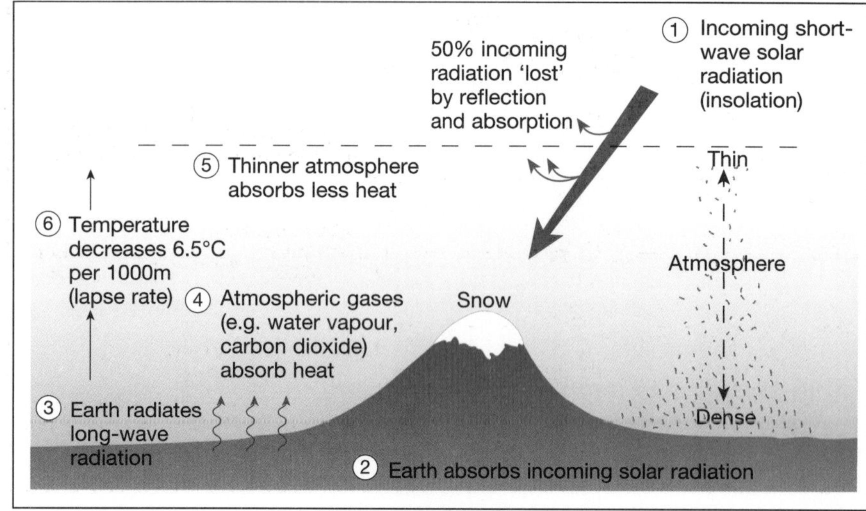

Fig 5.1 Heat received by insolation

- Solar radiation or **insolation**, is the main source of heat energy to the atmosphere.
- The atmosphere is warmed mainly from **long wave radiation** emitted by the Earth.
- Temperature therefore decreases with height at an average rate of 6.5°C per 1000m (**lapse rate**).

Air pressure

- Air pressure is the weight of the atmosphere at the Earth's surface.
- It is recorded using an **aneroid barometer**.
- It is measured in **millibars (mb)**, reduced to sea level (average pressure at sea level is 1013mb).
- On maps, differences in pressure are shown by **isobars**, which are lines joining points of equal pressure.
- Air pressure varies with:
  - temperature: warmer, lighter air is forced to rise creating areas of low pressure; heavier, colder air descends creating areas of high pressure
  - height – as the atmosphere becomes thinner, pressure decreases.

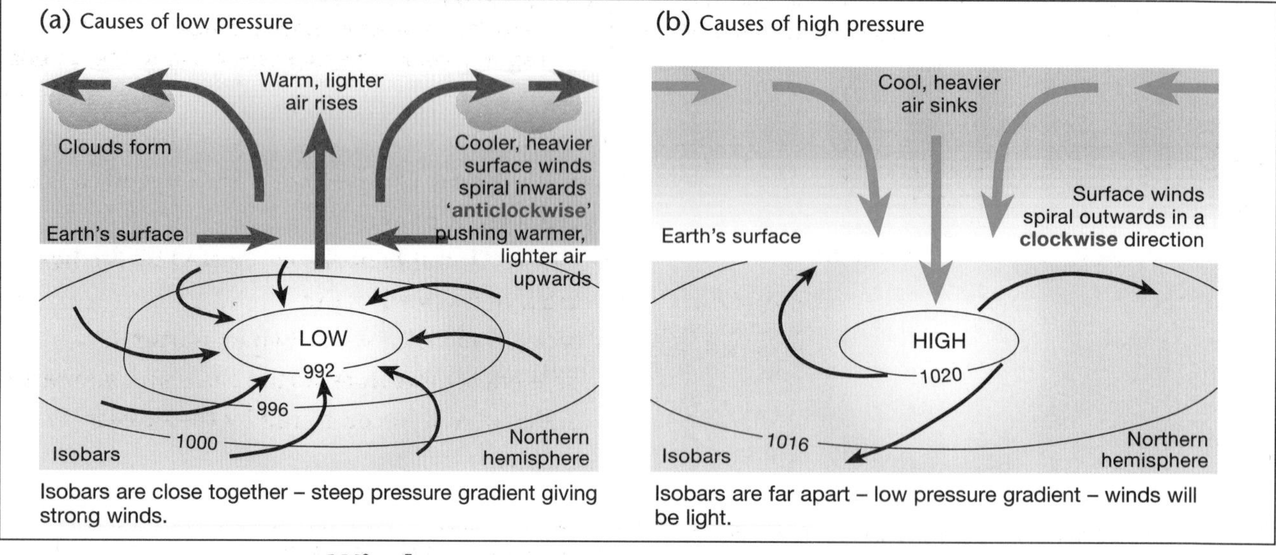

Fig 5.2 Causes of air pressure

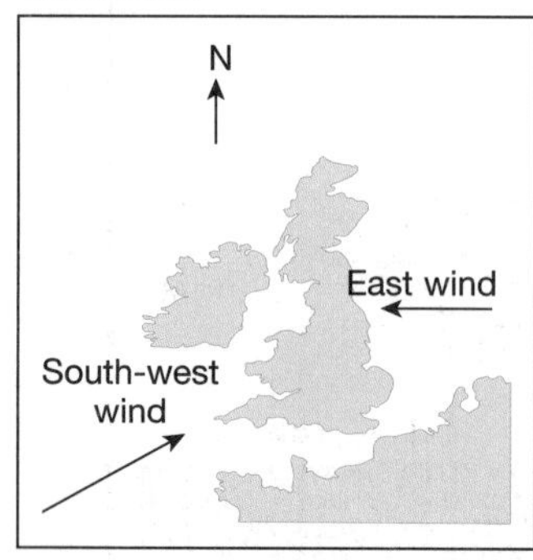

Fig 5.3 Giving wind directions

Wind

- Wind is the horizontal movement of air across the Earth's surface.
- It is recorded as:
  - direction using a **wind vane** (recorded as the direction from which the wind is blowing)
  - speed using an **anemometer**, measured in **knots, kilometres per hour** or on the **Beaufort Scale**.
- Wind is caused by differences in air pressure, the wind flows from high pressure (sinking) to low pressure (rising). The greater the pressure difference, the stronger the wind blows (see Fig 5.2a and 5.2b).

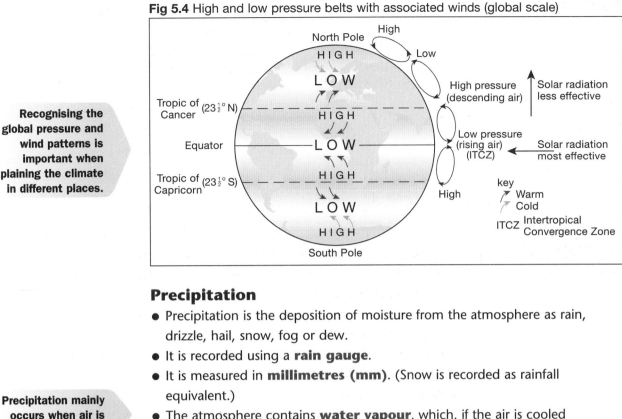

Recognising the global pressure and wind patterns is important when explaining the climate in different places.

Fig 5.4 High and low pressure belts with associated winds (global scale)

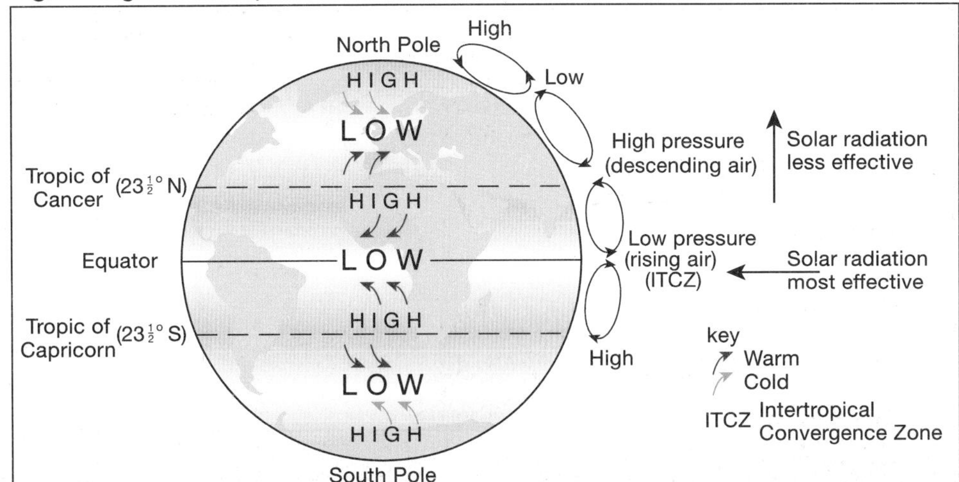

Precipitation

Precipitation mainly occurs when air is forced to rise.

- Precipitation is the deposition of moisture from the atmosphere as rain, drizzle, hail, snow, fog or dew.
- It is recorded using a **rain gauge**.
- It is measured in **millimetres (mm)**. (Snow is recorded as rainfall equivalent.)
- The atmosphere contains **water vapour**, which, if the air is cooled sufficiently, will form visible water droplets.

Air cools and reaches saturation or **dewpoint** → Water vapour will **condense** to **liquid** → Visible water droplets form:
- **clouds**, if the air is forced to rise
- **fog** and **dew** if the air cools close to the ground

Fig 5.5 Types of rainfall (or precipitation)

A Relief rainfall, e.g. Atlantic

Scottish Highlands
Aberdeen
Condensation
Precipitation
Dewpoint
Warm, moist air rises and cools
Mountain
Dry air descends and warms – clear skies – low rainfall
Rain shadow
Sea

Types of rainfall:
- **relief** rainfall (orographic) – air is forced to rise over hills or mountains
- **convectional** rainfall – air, heated by the hot ground, is forced to rise
- **frontal** rainfall – when two air masses of different temperatures meet; the warm air mass is undercut and forced to rise by the heavier cold air.

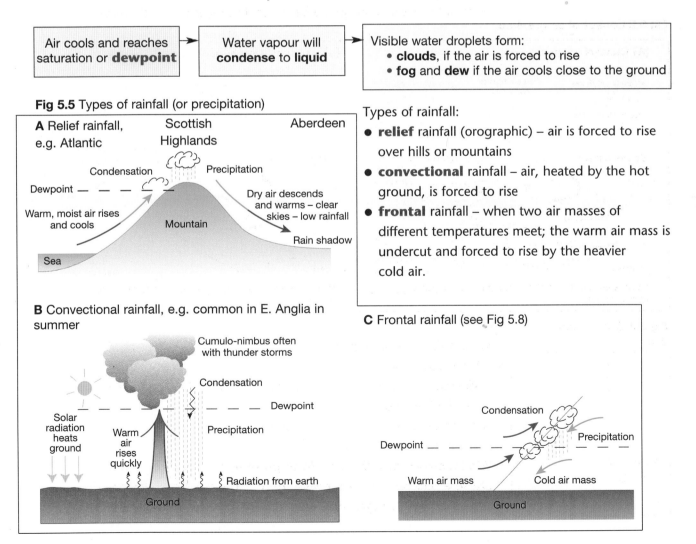

B Convectional rainfall, e.g. common in E. Anglia in summer

Cumulo-nimbus often with thunder storms
Condensation
Dewpoint
Solar radiation heats ground
Warm air rises quickly
Precipitation
Radiation from earth
Ground

C Frontal rainfall (see Fig 5.8)

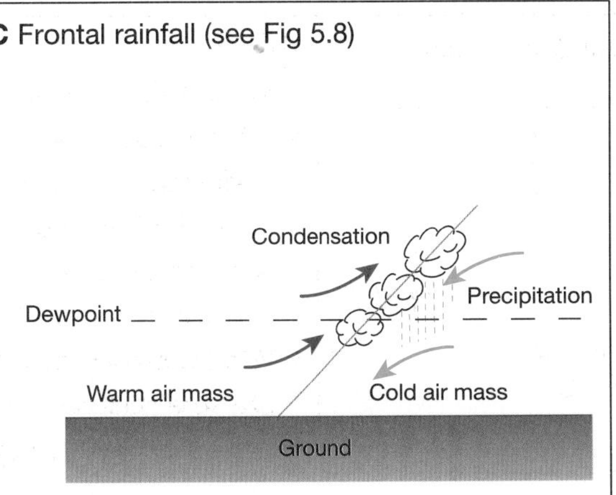

Fig 5.6 Types of clouds

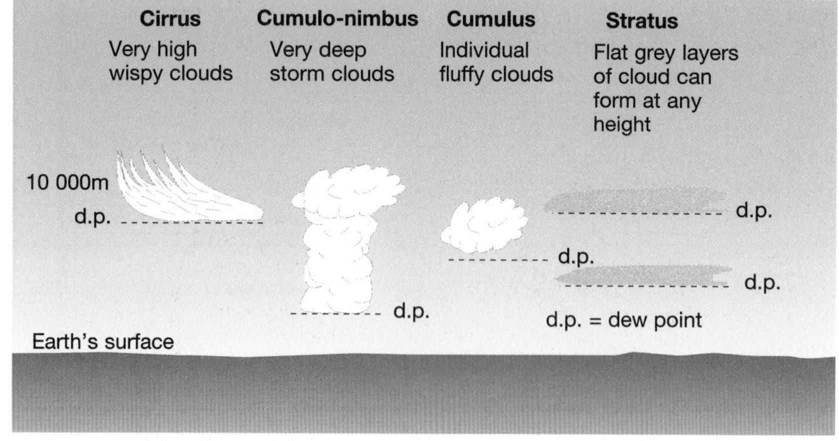

Clouds

- Clouds are visible masses of water droplets or ice crystals.
- They are recorded by observation; clouds are described by their shape and height.
- Cloud cover is measured in **oktas** (eighths of sky covered).

Sunshine

- Sunshine is recorded using a **sunshine recorder** and measured in **hours per day**.

Relative humidity (RH)

- Relative humidity is the amount of water vapour in the air.
- It is recorded by a **hygrometer** or **wet and dry thermometer**.
- It is measured as a percentage of the maximum water vapour that can be held by air of that temperature. It can be expressed in the following equation.

$$\frac{\text{water vapour present}}{\text{maximum water vapour that can be held by air of that temperature}} \times 100$$

- Warm air can hold more water vapour than cold air.
- If warm air is forced to rise and cool, RH will increase until the air becomes saturated, i.e. RH = 100% = **dewpoint**.

Fig 5.7 Weather map symbols

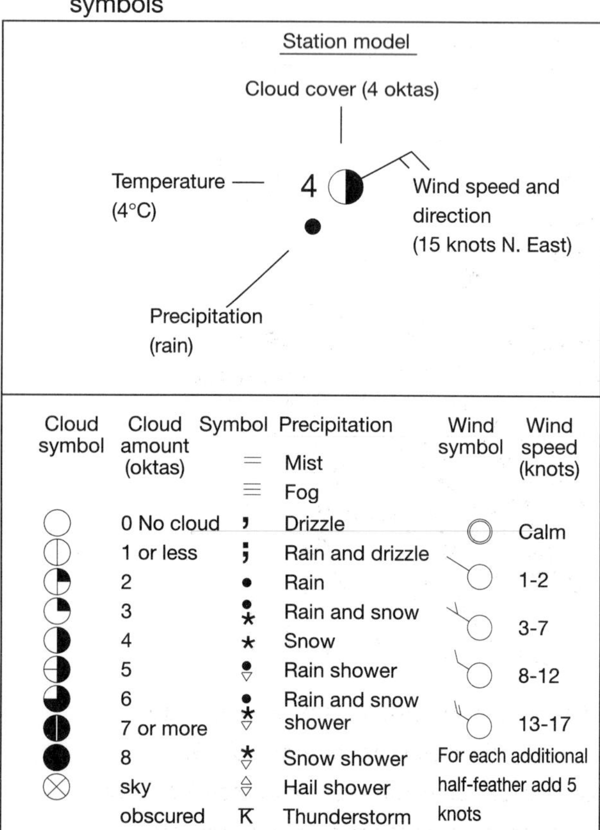

Weather maps or synoptic charts

KEY POINT **Weather maps or synoptic charts summarise the weather data for a particular time.**

- Data gathered from weather stations is represented as a group of symbols located at the site of the station.
- **Isobars** are drawn joining places of equal pressure.

Weather systems – depressions and anticyclones

Depressions

KEY POINT **Depressions are areas of low pressure, formed when a warm air mass meets a cold air mass. Fronts form at the boundaries of the two air masses.**

Fig 5.8A The formation of a depression

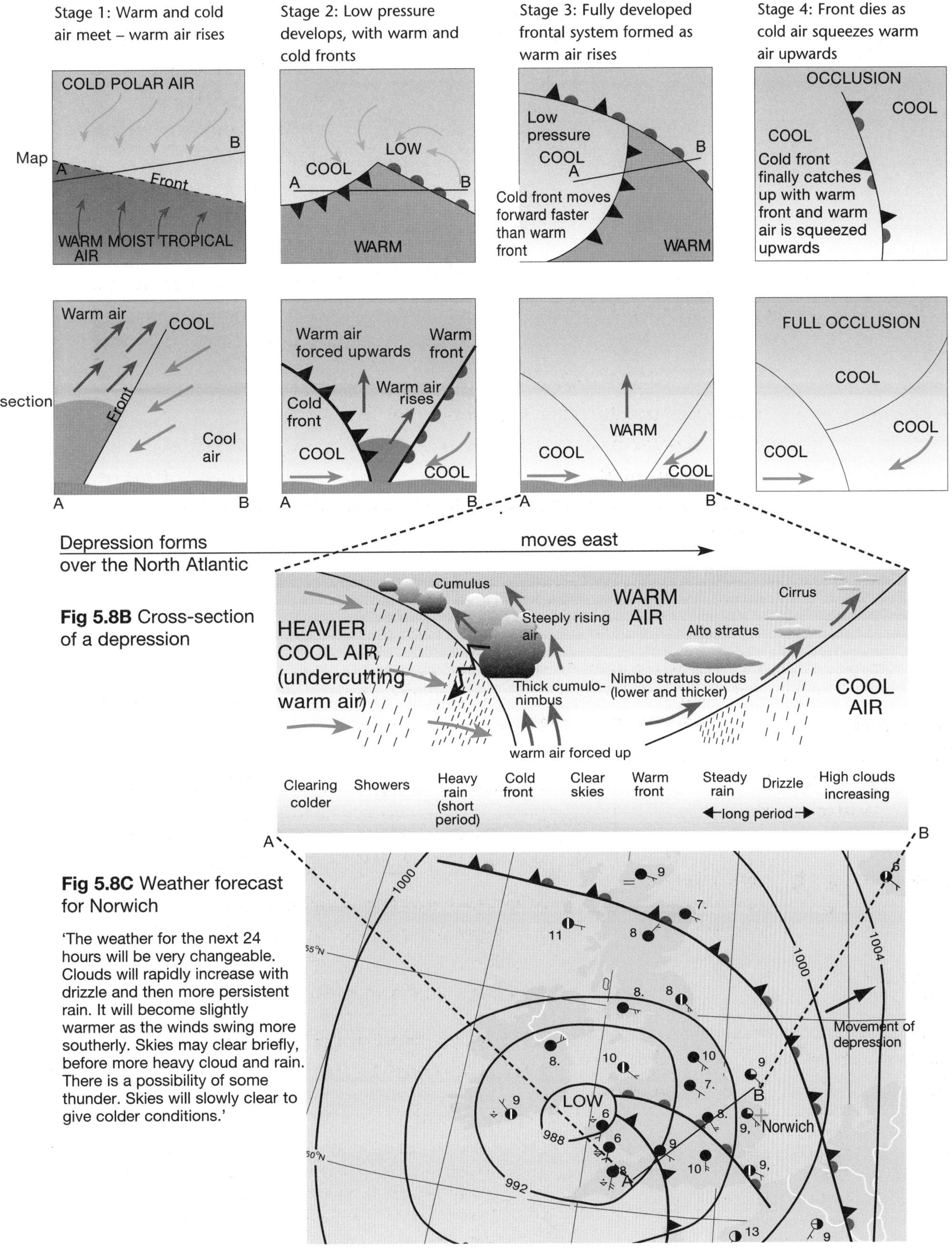

Fig 5.8B Cross-section of a depression

Fig 5.8C Weather forecast for Norwich

'The weather for the next 24 hours will be very changeable. Clouds will rapidly increase with drizzle and then more persistent rain. It will become slightly warmer as the winds swing more southerly. Skies may clear briefly, before more heavy cloud and rain. There is a possibility of some thunder. Skies will slowly clear to give colder conditions.'

- Depressions frequently form over the North Atlantic Ocean where warm, moist tropical air, moving north, meets cold polar air moving south.
- The depressions move east towards the UK, developing fronts (see Fig 5.8A).
- The cold air undercuts the warm air forming a cold front; the warm air is forced upwards.

Given a synoptic chart for the UK, be prepared to:
• <u>describe</u> the weather at a place given the weather symbols (Fig. 5.7)
• <u>explain</u> why it is raining close to the fronts that are shown (Fig 5.8b)
• <u>provide a forecast</u> for a place ahead of the depression (Fig. 5.8c)

- The warm air is pushed forward and forced to rise over the cold air ahead, forming a warm front.
- As warm air rises, clouds form and precipitation occurs along both fronts (see Fig 5.8B).
- Winds, in the Northern hemisphere, spiral inwards in an anti-clockwise direction. Winds will be strongest in deep depressions when the isobars are closer together.
- Meteorologists use their knowledge of depressions and fronts to make weather forecasts (see Fig 5.8C).
- Depressions are the main cause of the very changeable weather in the UK.

Anticyclones

KEY POINT **Anticyclones are areas of high pressure.**

It is important to be able to explain the weather (see also Fig 5.2b).

- In anticyclones, the air is **sinking** and warming.
- Warmer air can hold more water vapour, so clouds are unlikely to form.
- Anticyclones can become stationary, giving settled weather for several days or weeks in the UK.
- From 2004 to 2006, a series of anticyclones over the UK severely reduced rainfall. London and the SE received only 70% of their normal amount, threatening water supplies.

Fig 5.9 Anticyclones in summer and winter

Characteristic weather:

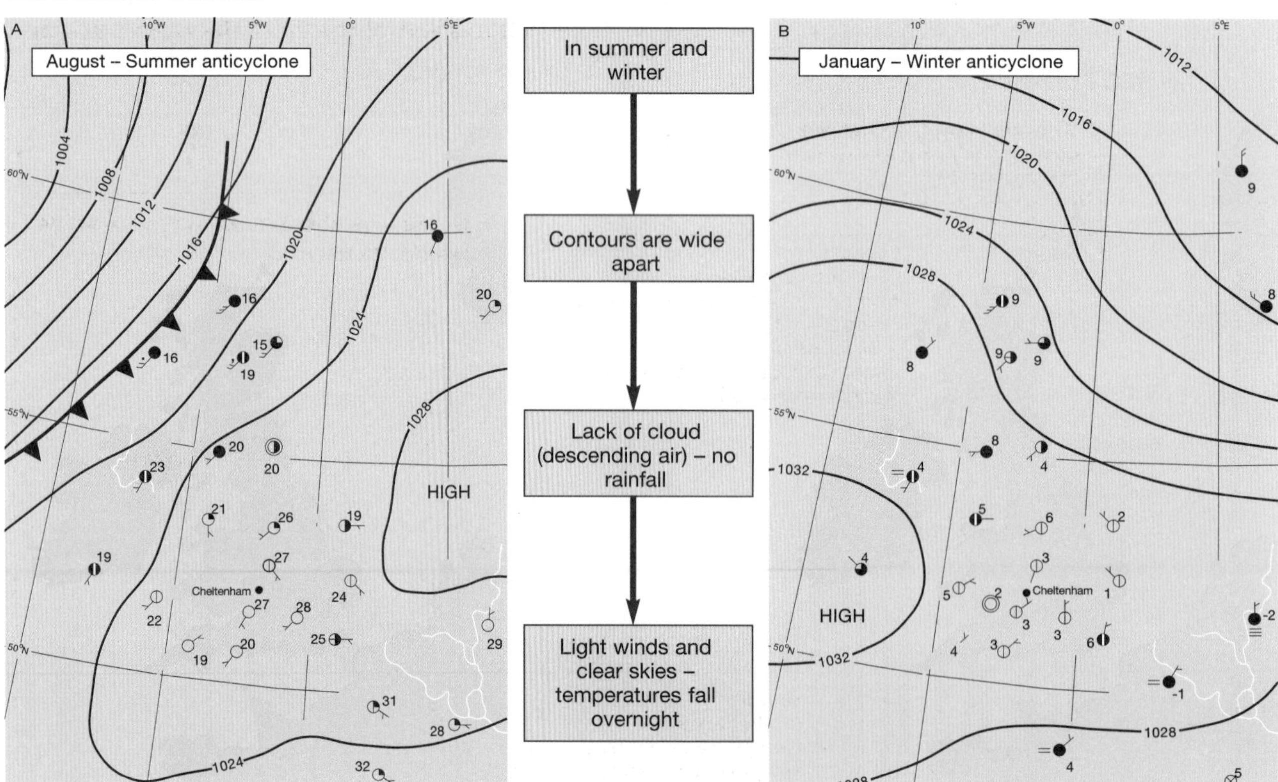

Average August temperature at Cheltenham: 27°C

Average January temperature at Cheltenham: 3°C

Clear skies – high solar heating – max. radiation – hot, fine summer weather. Misty evenings and nights, which clear quickly in the morning sun. Possible thunderstorms in the late afternoon in inland areas (see Fig 5.5b) May last for several weeks, giving a heatwave and possible drought (UK 1990).	Clear skies – low solar heating – max. radiation – cold crisp weather. Max. radiation through long winter nights – temperatures drop below 0°C – frosts and fogs, which may be slow to clear in the mornings. Possible snow showers.

- Winter anticyclones can cause **temperature inversions**, where a layer of warm air forms high above the ground and traps the clouds beneath it. During the winter the sun is not strong enough to 'burn off' the clouds. In January 2006, a temperature inversion reduced sunlight in London to 3.6 hours over nine days (25% normal). A gloomy start to the New Year!

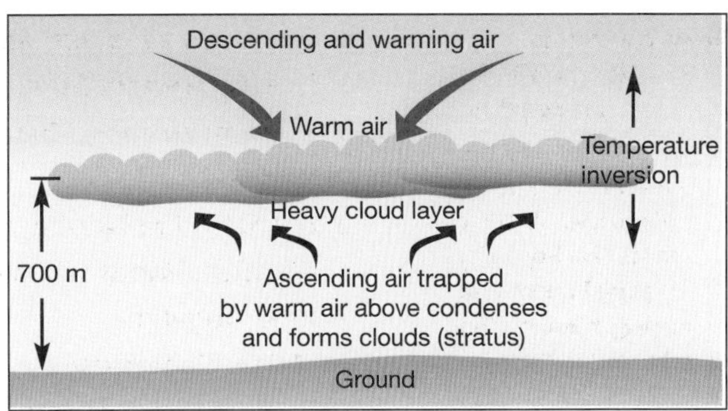

Fig 5.10 Temperature inversions

Climate

KEY POINT **Climate is the average weather conditions of a place or area, the result of weather data recorded over a long period of time (normally 30 years).**

Factors affecting climate

Fig 5.11 The effect of latitude on climate: insolation

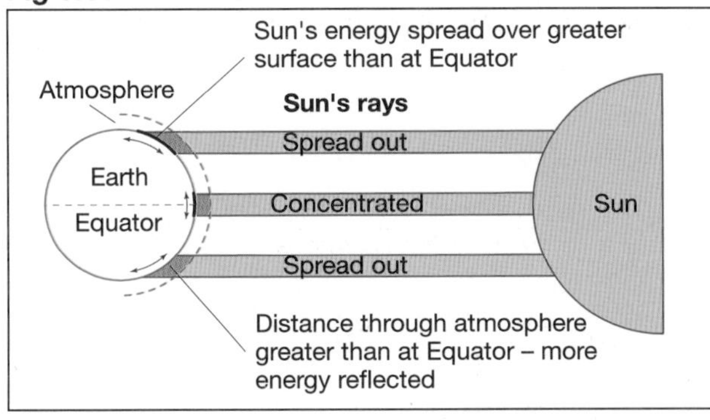

Latitude

With distance from the Equator:

- temperatures decrease as solar energy (insolation) is less concentrated towards the Poles
- seasonal differences in temperature are greater due to the effects of the tilt of the Earth's axis as it orbits the sun.

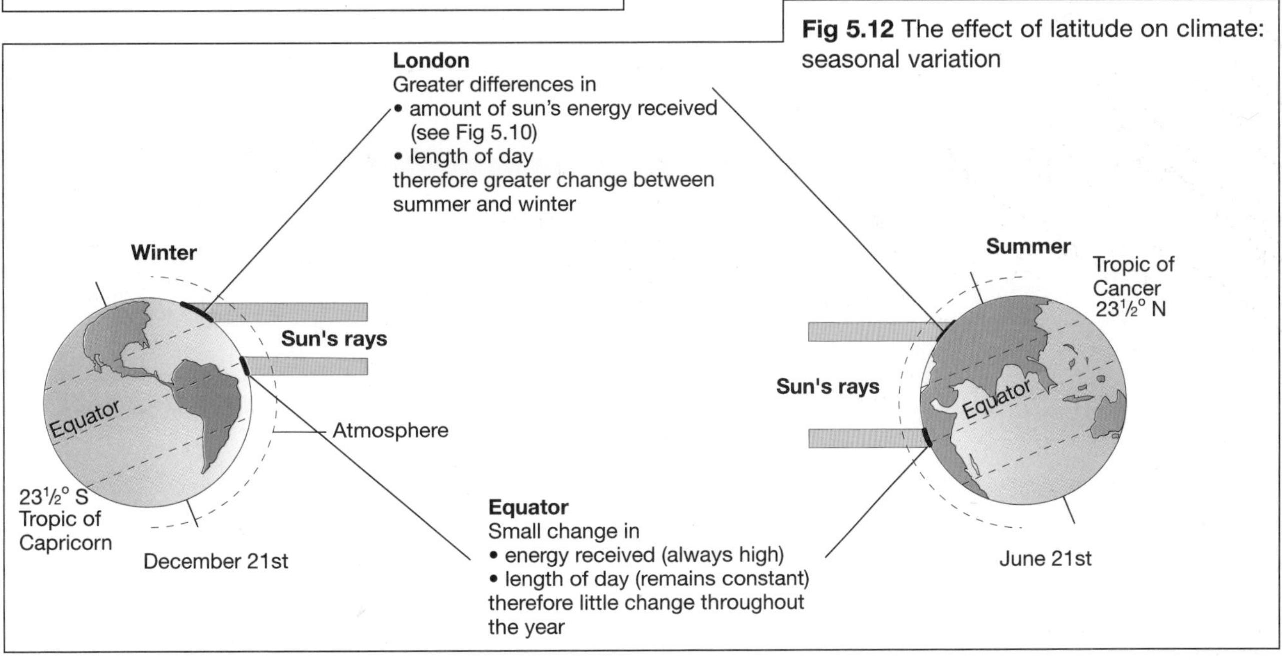

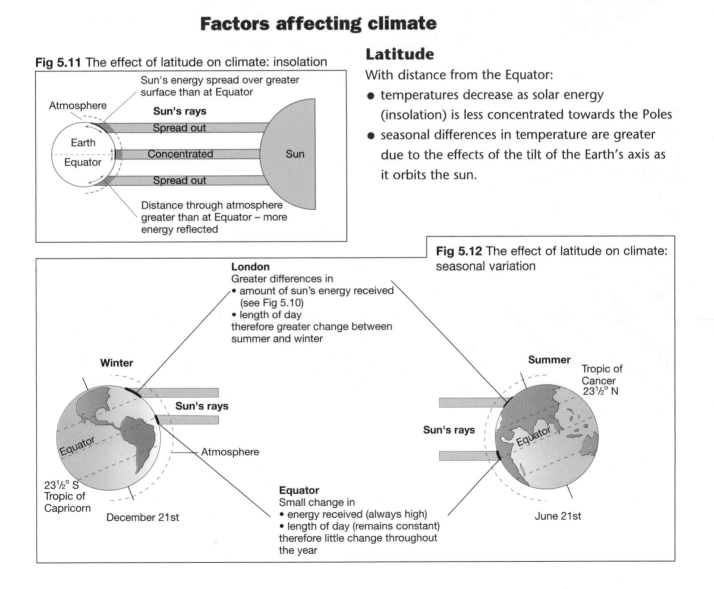

Fig 5.12 The effect of latitude on climate: seasonal variation

Altitude

Temperature falls, on average, 6.5°C for every 1000 metres in height (see Fig 5.1).

Fig 5.13 Altitude affects temperature, e.g. Scottish Highlands

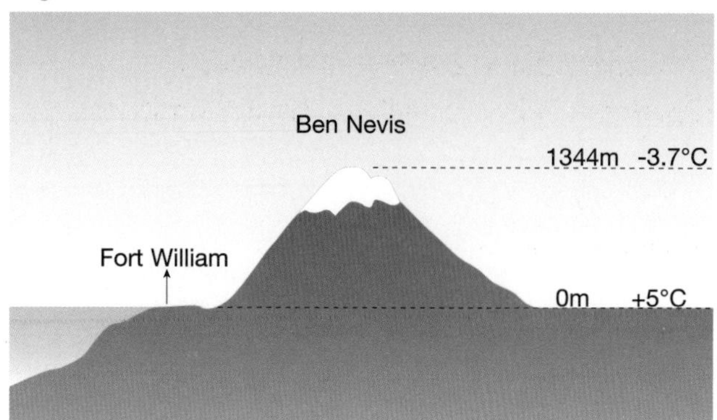

Distance from the sea

- The seas absorb and store heat to a greater depth than the land.
- The seas therefore warm up slowly in summer, storing heat which is slowly released through the winter. The land heats up more quickly but also cools more quickly.
- Coastal areas are kept cooler in summer and warmer in winter than land areas which are more distant from the sea.

Latitude, altitude, distance from the sea, prevailing winds and ocean currents are important factors affecting the climate. Refer to them when explaining the climate of any region.

- Islands and areas near the coast, such as the UK, have a **maritime** or **oceanic climate** with only a small range of temperature.
- Inland regions, such as central Germany, experience a more extreme **continental climate** with a large annual temperature range.
- The annual temperature range of two places at similar latitudes: e.g. Berlin 21°C (see Fig 5.16); London 14°C (see Fig 5.18).

Prevailing winds

- Prevailing winds are the most frequent winds affecting an area.
- Winds blowing from the sea are generally warm and moist at all times.
- Winds blowing out from large land masses are generally hot and dry in summer and very cold and dry in winter.

Ocean currents

- Warm and cold currents move large distances across oceans.
- Currents moving from the Equator towards polar areas bring warm conditions to coastal areas further north.
- Currents moving south from polar areas carry cold water south causing sea fogs particularly in summer.

Fig 5.14 The Labrador Current and the NAD

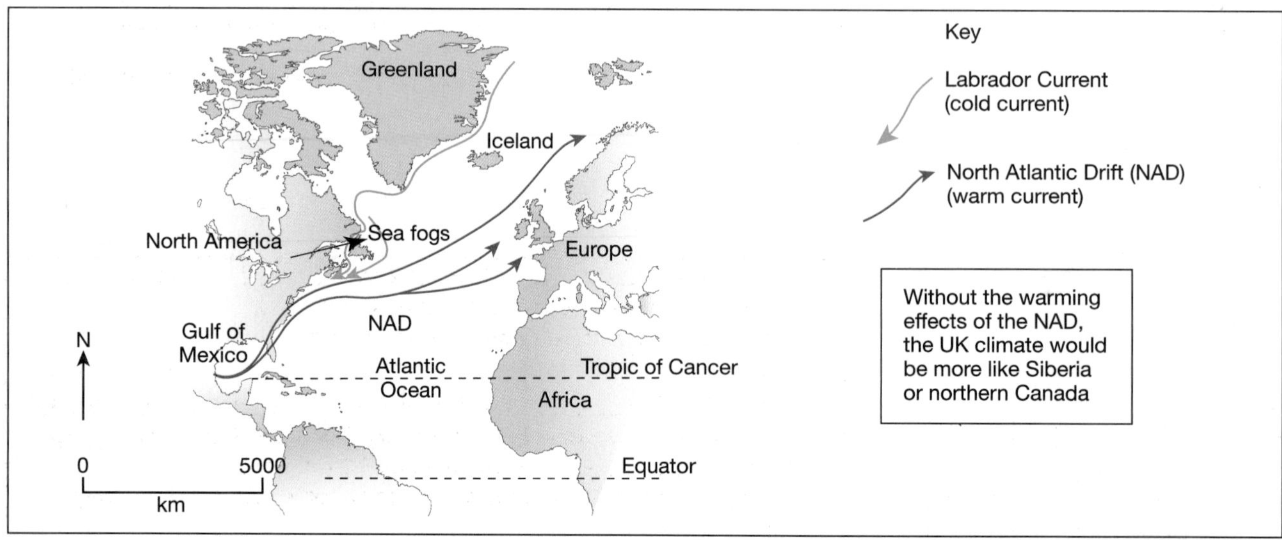

Climatic regions

AQA A

- At the global scale, the continents can be divided into climatic regions.
- These are extensive areas which have similar characteristics of temperature and precipitation.
- **Climate graphs** illustrate the main characteristics of each region.
- Maps show the extent and pattern of climatic regions. They illustrate the influence of:
  - latitude
  - ocean currents and prevailing winds
  - altitude.
- Within a climate region there will be many local variations of climate.

Fig 5.15 A climate map

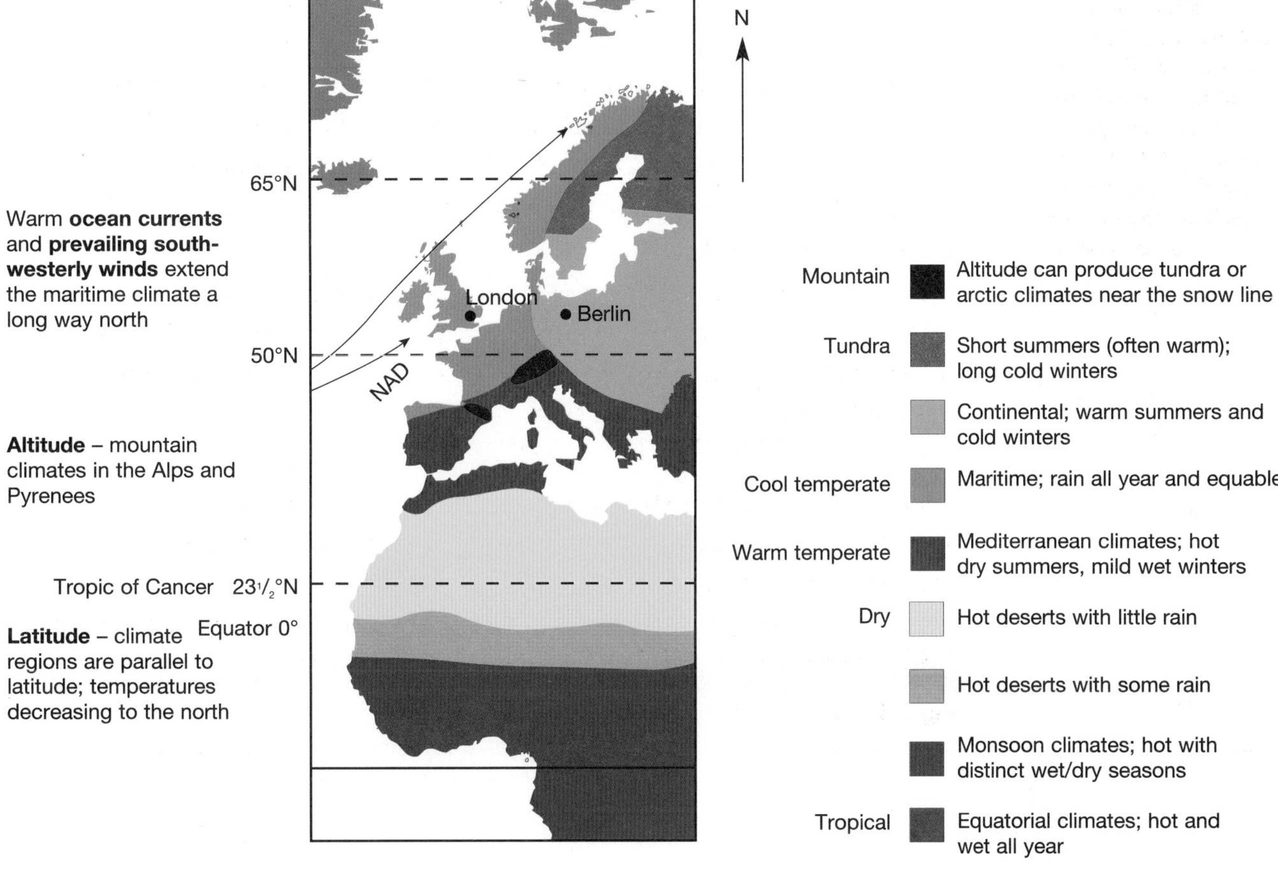

Fig 5.16

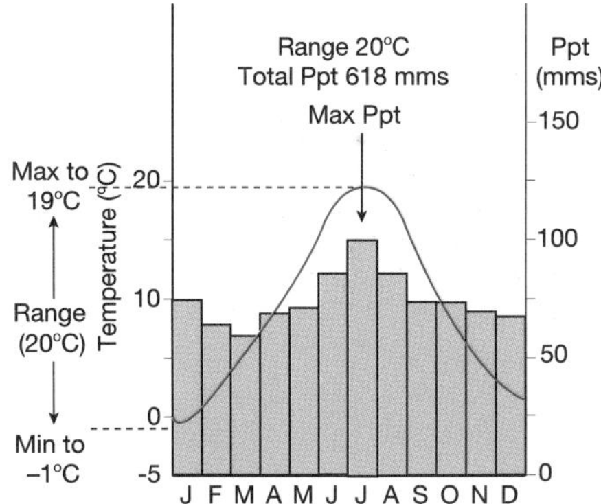

Climate graph for Berlin ($52\frac{1}{2}$°N – continental climate (see Fig 5.15 above)

- large temperature range (6°C greater than London – see Fig 5.18)
- short hot summers; very cold winters (Jan. −1°C)
- precipitation all year with summer maximum

Explanation:

- distant from influence of sea; land heats up and cools down quickly
- easterly winds in winter bring cold, dry, polar continental air – clear skies, hard frosts – heavy winter snowfalls common
- strong summer heating results in convectional storms and heavy rainfall

Weather and climate patterns in the UK

Fig 5.17A Seasonal temperature changes in the UK

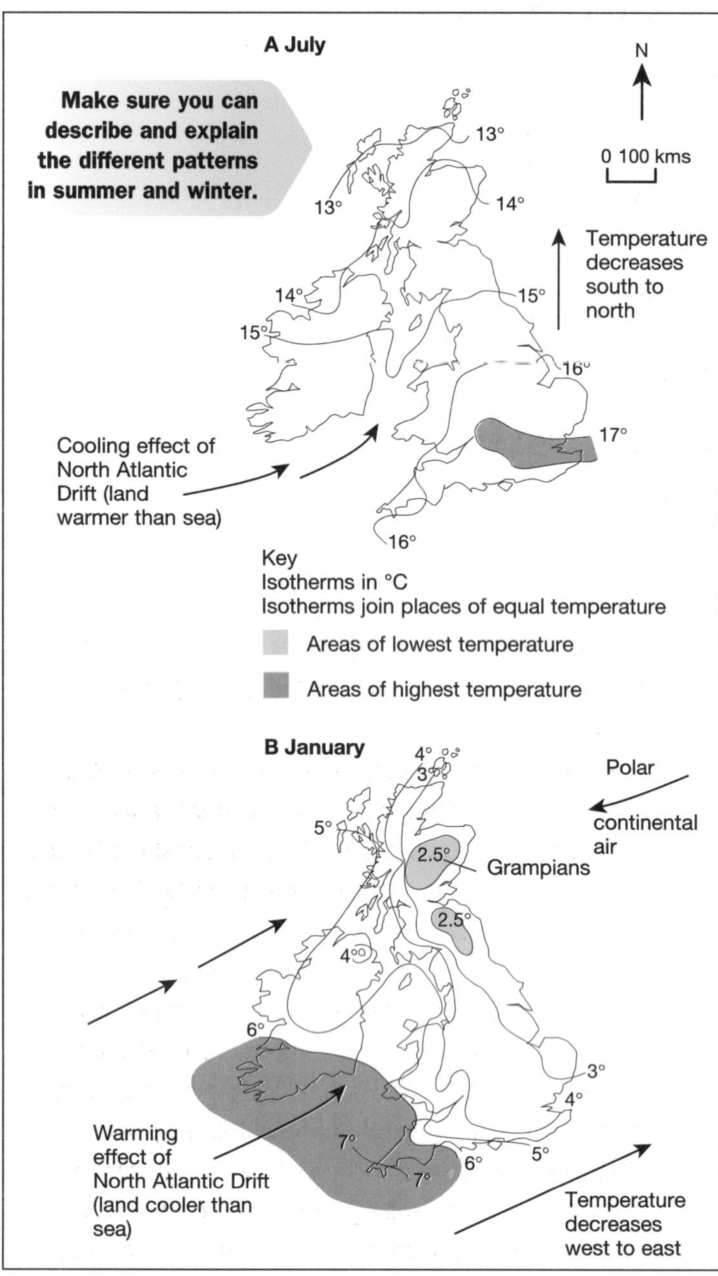

Fig 5.17B Precipitation patterns in the UK

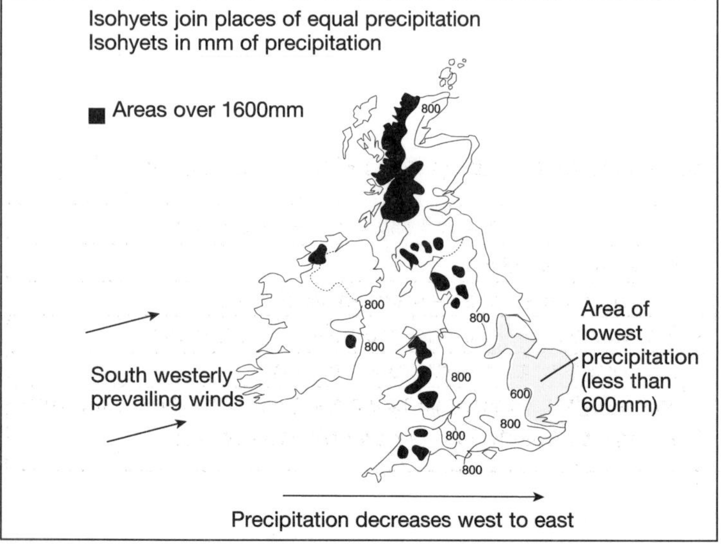

- Summer temperatures are warmer than winter temperatures (the south east is 13°C warmer).

Explanation: changes in solar energy due to tilt of Earth's axis as it orbits the sun (see Fig 5.12).

- In summer, temperatures are highest in the south-east (17°C) and coolest in northern Scotland (13°C).

Explanation: solar energy is less concentrated at higher latitudes (see Fig 5.11). Northern areas are more cloudy therefore have less sunshine.

- In winter, temperatures in the south-west (7°C) are 4°C warmer than in north-east Scotland.

Explanation: warming effect of the NAD (see Fig 5.14); cold polar air often reaches the north-east.

- Precipitation is highest in western areas (+1600mm) and decreases further east (less than 600mm).

Explanation: depressions approaching from the west bring large amounts of frontal precipitation (see Fig 5.8). Mountains in the west, e.g. Lake District, force south-westerly prevailing winds to rise resulting in relief rainfall; drier eastern areas are in the rain shadow (see Fig 5.5A).

- In western areas, most precipitation falls in the winter; in eastern areas more rain falls in the summer and autumn.

Explanation: there are more depressions in winter; summer heating in eastern areas results in convectional storms (see Fig 5.5B).

Contrasting climates

Fig 5.18 Climate graph for London (51½°N): cool, temperate maritime climate

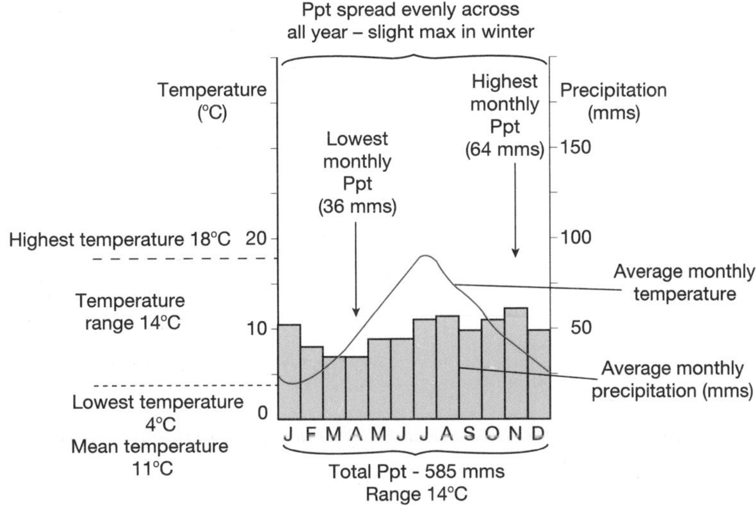

Fig 5.19 Climate graph for Uaupes (1°S): tropical equatorial climate (see Fig 5.15).

Find the location of Uaupes in your atlas.

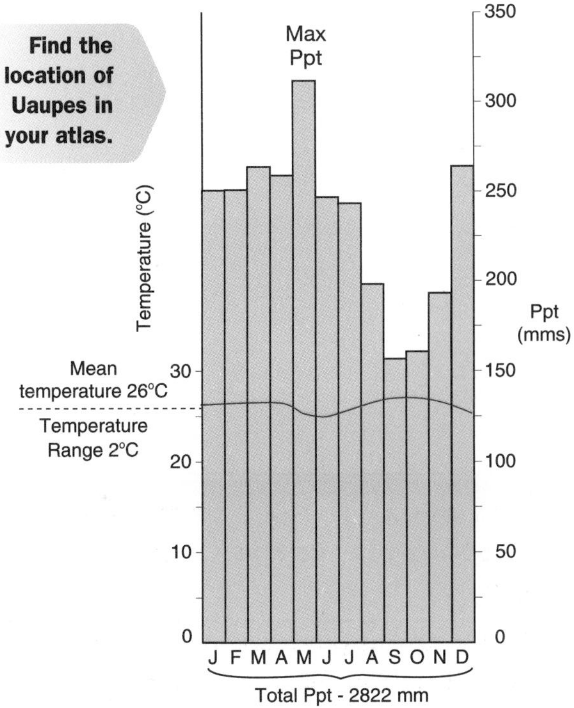

Uaupes, north west Brazil

- This climate occurs between 5°N and 5°S of the Equator.
- There are high temperatures all year, average 26°C. **Explanation**: close to the Equator, solar radiation is very high all year (see Fig 5.11); high sunshine hours.
- There is a very small temperature range (2°C) with no summer or winter seasons. **Explanation**: minimal effect of tilt of the Earth's axis (see Fig 5.12).
- There is a large day–night variation in temperatures (diurnal) (10°C). **Explanation**: clear skies at night – Earth cools rapidly (radiation) – early morning mists.
- There is very high precipitation (2822mm), over 150mm every month, and no wet and dry seasons. **Explanation**: high solar heating all day results in strong convection currents and clouds building up, giving heavy rain and thunderstorms every afternoon (see Fig 5.5B).

Table 5.1A Comparison of Uaupes with London – climate characteristics (refer to Figs 5.18 and 5.19)

	London	Uaupes	Uaupes compared with London
Mean temperature	11°C	26°C	More than double
Temperature range	14°C	2°C	Seven times smaller, no seasonal change
Highest temperature	18°C	27°C	9°C warmer all year than the summer in London
Lowest temperature	4°C	25°C	21°C warmer than January in London
Total precipitation	585mm	2822mm	Much higher, nearly five times as great; has more rain from January to March than London receives all year

Table 5.1B Comparison of Uaupes with London – vegetation and human activity

	London	Uaupes
Soils and vegetation	Deep fertile soil; deciduous woodland (see page 86)	Very deep but poor soils, particularly if trees are removed; tropical rainforest (see page 90)
Agriculture	Large commercial, intensive arable farms; capital intensive	Shifting cultivation, subsistence; minimum investment; commercial logging; cattle rearing after forest clearance
Population	Densely populated; very urbanised	Low density; semi-nomadic tribes

Weather hazards – tropical storms

AQA A AQA B AQA C

KEY POINT

Tropical storms are large low pressure systems which develop over oceans in the Tropics during the summer. They move north and south away from the Equator, along similar tracks. They are known in different parts of the world as cyclones, hurricanes, typhoons and willy-willies.

Tropical storms are hazardous particularly if they reach land because of:

- heavy rainfall (200mm per day) which can cause flooding and landslides damaging property and communications
- storm surges – low pressure causes the sea level to rise; large waves form in shallow coastal waters (+5m high). Low-lying coasts are very vulnerable, causing severe loss of life in highly populated areas
- high winds – often exceeding 200kph, cause severe damage to buildings, crops and electric wires.

Research the effects of Hurricane Katrina 2005 and the storm surge that affected New Orleans and the coast of Texas.

Tropical storms form:

- over warm, tropical oceans with a water temperature of 27°C
- where high solar radiation causes rapid evaporation.

Stages in the development of a tropical storm

Fig 5.20 Cross-section through a tropical storm

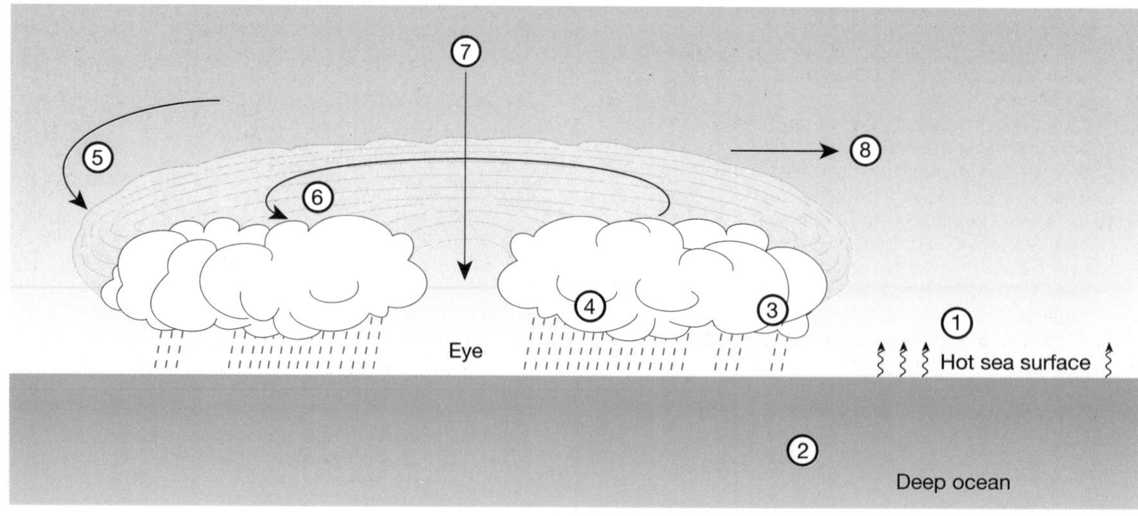

1 High solar heating; strong evaporation
2 Deep ocean; hot surface water (27°C)
3 Warm, very moist air rises quickly, cools and condenses
4 Very deep cumulo-nimbus clouds form; heavy rainfall
5 Low pressure develops; high winds as air is sucked into the depression
6 Rotation of the Earth encourages violent rotating winds around the central 'eye' (anticlockwise in northern hemisphere)
7 Central eye – descending air, clear skies, calm conditions
8 Whole system moves forward very fast in the prevailing winds (15-25kph)

Use the sequence shown, 1–8, to develop a logical explanation.

Effects of tropical storms

MEDC case study: Hurricane Floyd

Fig 5.21 Hurricane Floyd menaces Georgia and the Carolinas

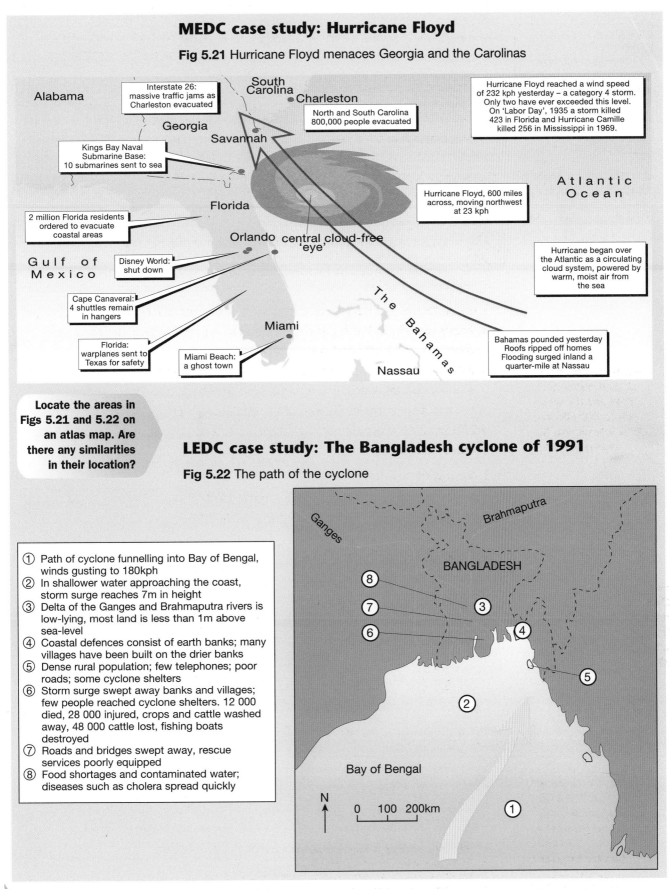

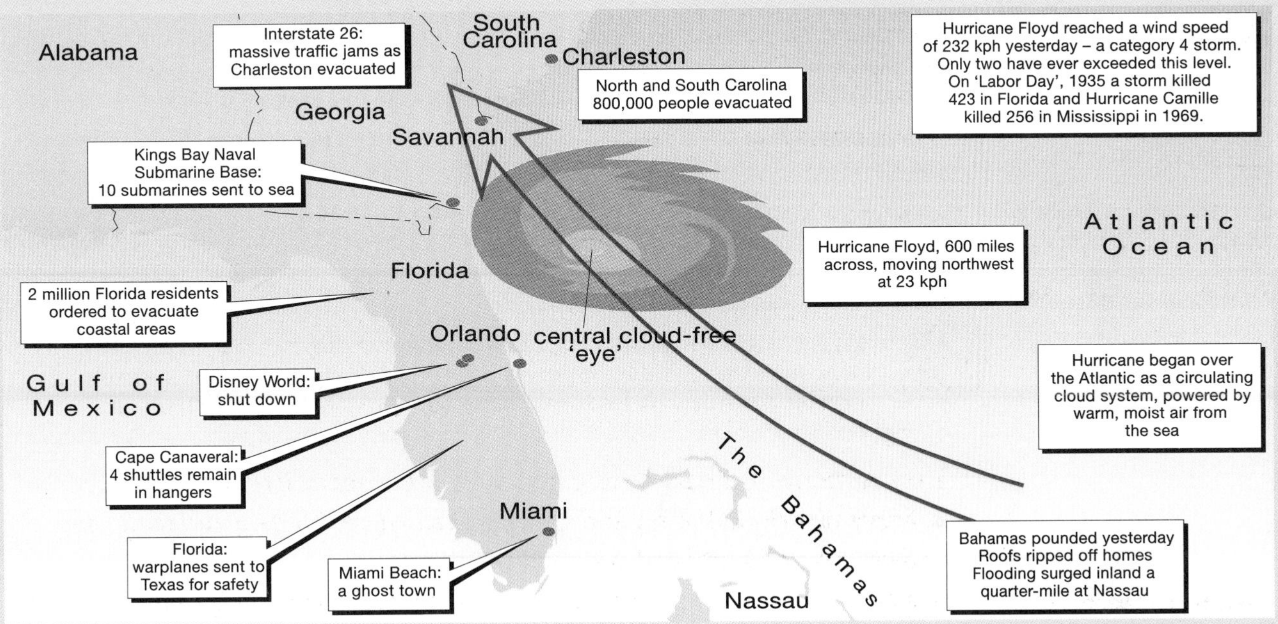

Locate the areas in Figs 5.21 and 5.22 on an atlas map. Are there any similarities in their location?

LEDC case study: The Bangladesh cyclone of 1991

Fig 5.22 The path of the cyclone

① Path of cyclone funnelling into Bay of Bengal, winds gusting to 180kph
② In shallower water approaching the coast, storm surge reaches 7m in height
③ Delta of the Ganges and Brahmaputra rivers is low-lying, most land is less than 1m above sea-level
④ Coastal defences consist of earth banks; many villages have been built on the drier banks
⑤ Dense rural population; few telephones; poor roads; some cyclone shelters
⑥ Storm surge swept away banks and villages; few people reached cyclone shelters. 12 000 died, 28 000 injured, crops and cattle washed away, 48 000 cattle lost, fishing boats destroyed
⑦ Roads and bridges swept away, rescue services poorly equipped
⑧ Food shortages and contaminated water; diseases such as cholera spread quickly

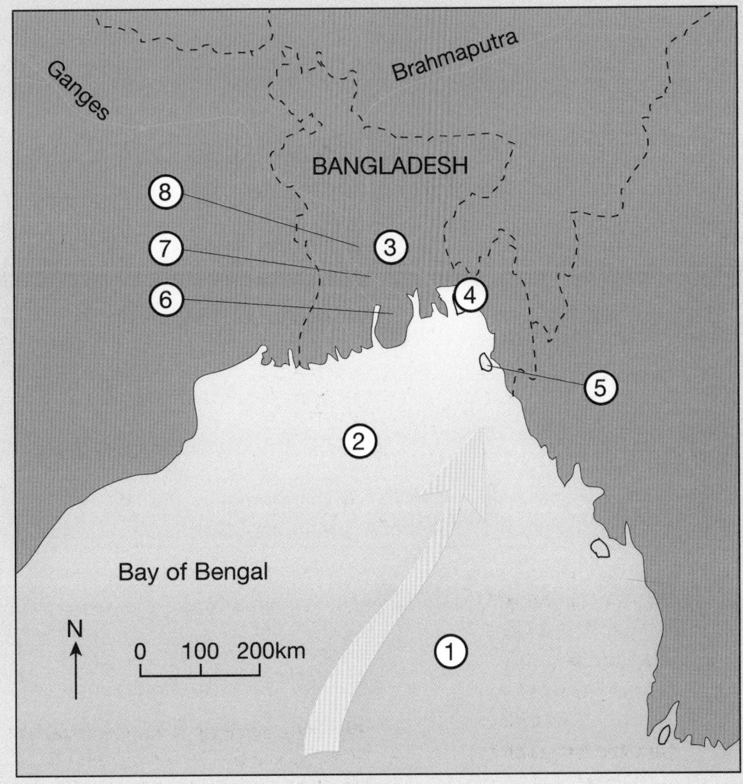

Prediction and protection

Keep up-to-date by keeping newspaper reports on current hazards and disasters.

Be prepared to compare an MEDC with an LEDC in terms of underlying problems due to the difference in wealth. Compare the GNP of Bangladesh with that of the USA.

MEDCs:

- monitor and track storms by satellite and aircraft
- issue storm warnings at least 12 hours in advance
- prepare evacuation plans, e.g. designate large sports stadiums as refuges
- people prepare by:
  - building personal storm shelters
  - boarding-up windows
  - having adequate insurance cover
  - storing food and clean water supplies
- train emergency services

LEDCs:

- build cyclone shelters (£80 000 each), capable of safely housing a large number of families
- raise and strengthen earth banks along coasts and rivers
- plant mangrove trees along coasts to build up silt and absorb storm surges
- educate people regarding risk
- obtain aid from MEDCs

Drought and desertification

AQA C

Drought

A drought is a long period of time without significant rainfall.

Factors affecting the severity of droughts

- Climatic factors
  - Past rainfall will affect the amounts of water in surface stores (rivers, lakes), and soil and groundwater (aquifer) stores (see Fig 2.2, page 22).
  - Rock type: permeable rocks, e.g. chalk, are able to store more water than impermeable rocks, e.g. clay – rapid run-off into rivers (see Fig 1.11).
  - Temperature will affect the amount of water lost by evaporation.
- Human factors
  - The size of a country's population will affect the demand for water.
  - The economic wealth of a country will affect demand and its ability to cope with the problems (alternative water supplies).
  - Land-use: demands from agriculture (irrigation) and industry.

Causes of droughts in the UK

An official drought in the UK is defined as a period of at least 15 consecutive days with 0.2mm of rain or less.

Winter rainfall in 2005–2006 in SE England was the lowest since the 1930s. Research the outcomes in summer 2006.

- Lower than average rainfall in winter and spring.
- Summer anticyclones (high pressure systems) that become stationary over the UK for several weeks (see Fig 5.9) bringing clear, calm weather.
- Higher than average demand for water in hot weather.

Impact of droughts in MEDCs

- Water rationing by using street standpipes or tanker supplies (UK 1976 and 1995).
- Reduced flows in rivers concentrate pollutants and reduce oxygen in water causing fish and aquatic plants to suffer or die.

- Trees and plants become very vulnerable, particularly in urban areas.
- Reduction in yields of crops and grass, may lead to increased food prices.
- Damage to housing from shrinkage and cracking of clay soils.
- Increased fire risk in forests and on heathlands, e.g. New Forest, Hampshire. Can threaten and destroy property, e.g. California 2000; Sydney 2005.
- Droughts are often associated with heatwaves (high pressure – see Fig 5.9) with risks of air pollution in urban areas (see Fig 5.28).

Compare the impact of drought in MEDCs with that in LEDCs.

Managing the effects of drought in MEDCs

Short term

- Water conservation measures, e.g. banning hosepipes/washing cars.
- Educate people in saving water, e.g. *'Everyone in England and Wales could save four litres a day by turning off the tap while brushing their teeth, which would be enough to supply more than 600 000 homes every day'.*

Medium term

- Compulsory water metres (Kent UK 2006)

Long term

- Building new reservoirs.
- Establishing a national water supply network, transferring water from areas of high rainfall, e.g. NW England, to areas of high demand and low rainfall, e.g. SE England.

Causes of droughts in LEDCs

Case study: The Sahel

The Sahel is a belt of semi-arid land that stretches across Africa from Mauritania in the west to Somalia in the east. Since 1968 the area has suffered from many severe droughts.

Fig 5.23 Drought in the Sahel (see Fig. 5.4) – failure of the summer rains

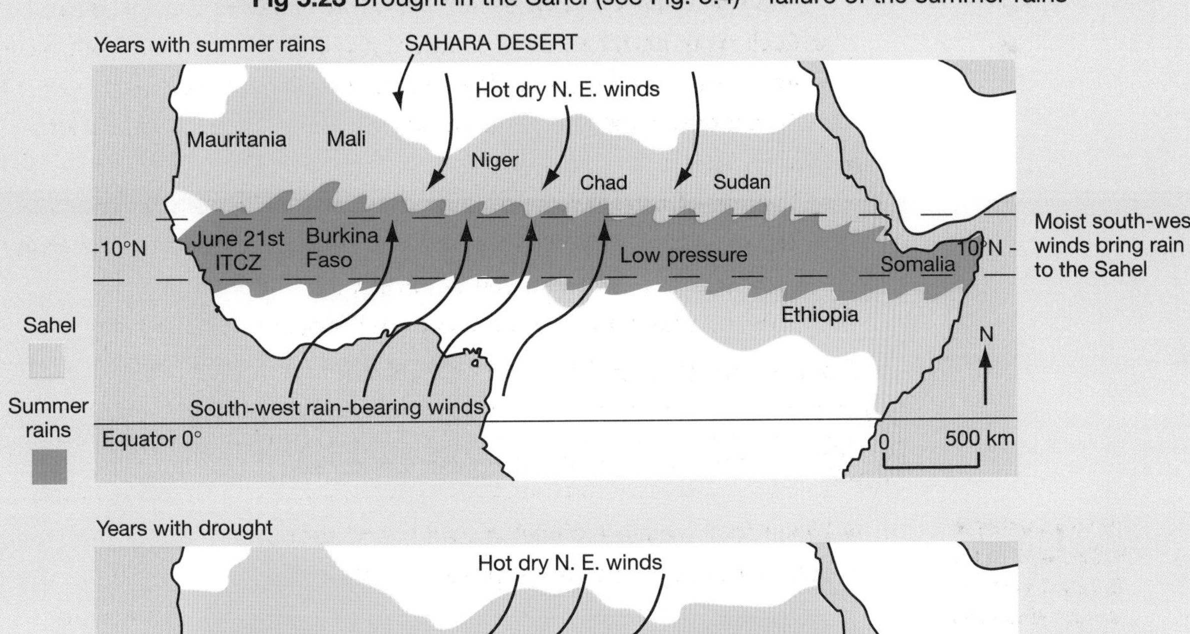

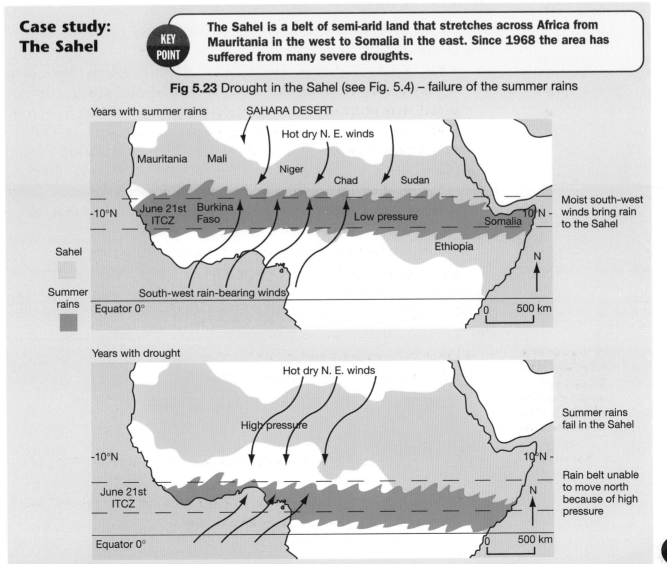

Many scientists blame climate change for disturbing the global circulation of winds and ocean currents and causing the rains to fail in the Sahel. Other scientists consider the droughts to be only short-term fluctuations of climate.

- With less rainfall, trees and grassland cannot survive and becomes scrub vegetation.
- Less vegetation gives less transpiration, so a smaller amount of water vapour is available and therefore there is less likelihood of rain.

Desertification

Desertification is the process by which land becomes desert.

Desertification is also spreading north of the Sahara, in countries such as Algeria, Spain, Italy and Turkey.

Drought and human activities have led to the Sahel becoming more like a desert, with:

- thin, dry, sandy soils,
- bare rock where the soils have been blown away (soil erosion)
- sparse vegetation.

The processes of desertification

Fig 5.24 Desertification in the Sahel

NOTE: Desertification is the result of both natural and human causes.

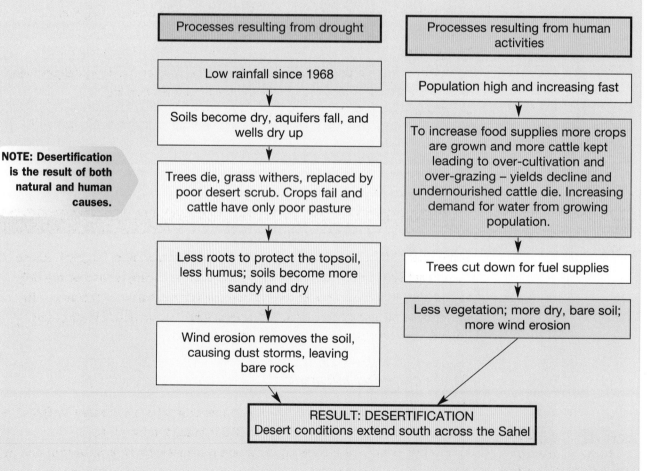

Processes resulting from drought	Processes resulting from human activities
Low rainfall since 1968	Population high and increasing fast
↓ Soils become dry, aquifers fall, and wells dry up	↓ To increase food supplies more crops are grown and more cattle kept leading to over-cultivation and over-grazing – yields decline and undernourished cattle die. Increasing demand for water from growing population.
↓ Trees die, grass withers, replaced by poor desert scrub. Crops fail and cattle have only poor pasture	↓ Trees cut down for fuel supplies
↓ Less roots to protect the topsoil, less humus; soils become more sandy and dry	↓ Less vegetation; more dry, bare soil; more wind erosion
↓ Wind erosion removes the soil, causing dust storms, leaving bare rock	

↓

RESULT: DESERTIFICATION
Desert conditions extend south across the Sahel

Impact of drought and desertification in LEDCs

- Food and water shortages lead to malnutrition, famine, disease and high death rates.
- Large numbers of people become dependent on food aid programmes (see page 178).
- There is migration of people
  - from rural to areas causing over-population in towns
  - to refugee camps.

Population growth continues to outstrip food supplies.

Civil wars as in Eritrea, Ethiopia and Sudan prevent aid reaching stricken areas, and cause mass migrations to refugee camps.

Solutions

Short term

- Provide food aid and water supplies to prevent suffering.
- Conserve water in local, small-scale schemes, e.g. Burkina Faso – building low stone walls across shallow slopes.

Medium term

- Improve farming methods by encouraging schemes using sustainable, appropriate technologies, e.g. small-scale, locally made tools – not tractors (see page 179).
- Provide drought-resistant seed such as millet, e.g. northern Nigeria.
- Start tree-planting schemes to reduce soil erosion.

Long term

- Improve water supplies by building large reservoirs and drilling deeper wells.
- International action to reduce the causes of global warming.
- Long term sustainable aid (training, eduction).

Climate change

Some scientists believe that the global climates are changing, largely because of human activity. Other scientists believe the changes being measured are only fluctuations in the atmospheric conditions, which have always occurred – ice sheets have advanced across the UK several times in the last 200 000 years (long before human activity had significant effects).

Evidence for change

Articles in newspapers frequently discuss climate change when storms or flooding cause damage in the UK. Try to highlight the evidence for both sides of the argument.

- Rising global temperatures: average temperatures have increased by 0.6°C in the last 100 years – a further rise of 3°C is predicted by 2100.
- Extreme events are increasing: there are more intense tropical storms, and in the last few years the UK has had the driest, wettest and windiest periods of weather since records began.
- Changing ocean currents: currents in the South Pacific may be changing and causing droughts and forest fires in Indonesia and the loss of important fishing grounds off Peru and Chile. The North Atlantic Drift (NAD) may be weakening (see Fig 5.14)

- Glaciers and ice sheets are shrinking; permanently frozen soils and rock in Siberia and on high mountains in the Alps are melting (increase in mudslides). Evidence from satellite images show that the Antartica ice-sheet is losing up to 36 cubic miles of ice each year (sea-level changes).
- Various birds, fish and insects found in warmer regions in Africa are spreading north into Europe, e.g. malarial mosquito.

The principal environmental problems are: **global warming**, **ozone layer depletion**, **photochemical smog** and **acid deposition**. These problems require international solutions and cannot be dealt with by one country alone.

Global warming

Frequently a subject of exam questions.

Global warming describes and explains the pattern of increasing global temperatures (see also pages 150–1.).

The greenhouse effect

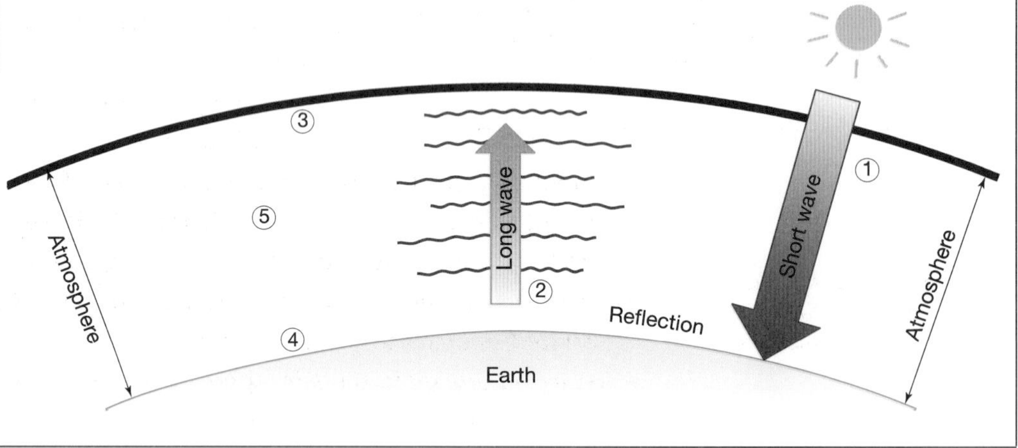

Fig 5.25 The greenhouse effect

1 Solar energy reaches the Earth as short-wave radiation
2 The Earth absorbs heat and radiates energy as long-wave radiation
3 Greenhouse gases, e.g. water vapour, carbon dioxide (CO_2), methane, absorb the long-wave radation and insulate the Earth
4 Without insulation, the Earth would suffer extremes of temperature and would be uninhabitable
5 Greenhouse gases occur naturally but human activities have resulted in rising concentrations, more heat is retained and the atmosphere is becoming warmer

Causes

- **Fossil fuels** – the burning of fossil fuels, such as coal and oil, releases CO_2 into the atmosphere. The rate at which fossil fuels are burnt has increased rapidly in the last 100 years. Energy consumption in power stations, for domestic heating and in cars and lorries has increased by 50% in the last ten years.
- **Deforestation** – trees convert CO_2 into oxygen through photosynthesis. Most woodlands in temperate areas, such as the UK, have been chopped down. Now tropical rainforests are being felled. The burning of forests releases large amounts of CO_2 into the atmosphere.
- **Cattle farming** – cattle produce methane in their intestines; numbers of cattle have steadily increased as demand for food increases.
- **Waste tips** – decomposition releases large quantities of methane; commercial and domestic waste has increased enormously.

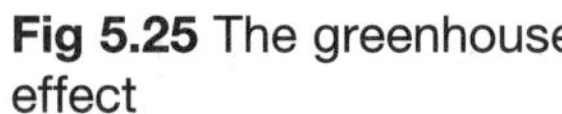

Do not confuse the greenhouse effect with holes in the ozone layer (see page 76).

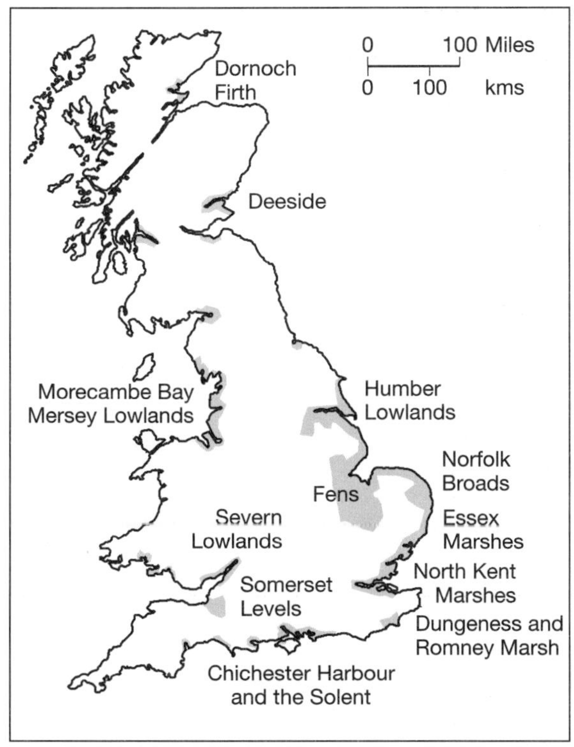

Fig 5.26 Areas at risk from sea-level rise

Consequences

- A rise in sea-level:
  - melting of ice caps and glaciers will release large amounts of water into the oceans
  - higher water temperatures will cause water to expand
  - flooding of low-lying coastal areas such as Bangladesh, Japan and the UK Fenlands, and submerging of islands in the Caribbean and Indian Ocean will produce many millions of refugees.
- Global circulation will be disturbed:
  - tropical areas are becoming drier and deserts expanding which will lead to widespread crop failures in central Africa, malnutrition and starvation
  - water shortages will increase particularly in Africa, northern China and India
  - spread of malarial mosquito in Africa, USA and Europe
  - decreasing crop yields in South America, China, Africa and India
  - more temperate conditions in Canada and Siberia will improve agriculture.
- The melting of **permafrost** in Siberia will release large amounts of methane into the atmosphere and further increase global warming.
- Increased evaporation will increase amounts of water vapour in atmosphere and further increase global warming.
- Increased energy in the atmosphere will increase the likelihood of **extreme events** such as tropical storms.

A few areas benefit but most will have problems.

Strategies to reduce global warming

- Reduce emissions of greenhouse gases:
  - develop **alternative forms of energy**, e.g. solar, tidal, wind, hydroelectricity (see page 145)
  - switch to nuclear power (see page 144)
  - develop **alternative energy sources** for road transport (electric cars)
  - support LEDCs in using alternative forms of energy
  - agree international commitments to reduce emissions and limit pollution, e.g. **Earth Summits**
    - 1992 Rio de Janeiro (Agenda 21, 118 countries committed)
    - 1997 Kyoto (5% cut in emissions by all MEDCs)
    - 2000 The Hague (USA is generally opposed to reductions, yet USA produces 5.5 million tonnes of CO_2 per year (1996), five times UK emissions and 230 times Bangladesh emissions)
    - 2005 Kyoto agreement
- Reduce **deforestation** and develop sustainable policies such as **selective cutting** (rainforest) and **replanting** (UK National Forest e.g. Leicestershire).
- Reduce population growth and reduce energy consumption.

Renewable sources of energy.

Keep up-to-date with 'new summits' which try to reduce pollution.

Why are some countries unwilling to reduce their emissions of CO_2?

LEDCs will be more affected by drought and famine. How will these countries afford to pay for the research and new developments?

Strategies to reduce impact of global warming

- Defend and protect low-lying coastal areas from flooding.
- Improve water supplies and use them more efficiently, particularly in areas that are becoming drier.
- Encourage research into developing crops to withstand drought and diseases.
- Develop more efficient ways of predicting and preparing for extreme weather events such as tropical storms.

Ozone layer depletion

Ozone (O_3) forms a protective layer of gas around the Earth that absorbs and filters ultra-violet (UV) radiation from the sun.

Do not give this as a reason for global warming.

Damage to the ozone layer

- In 1986 scientists in Antarctica discovered a 'hole' in the ozone layer.
- Greater amounts of **UV radiation** are reaching the Earth.
- Measurements in 1998 showed that the hole had grown larger and amounts of **UV radiation** had increased in Australia and Argentina.
- A further hole was discovered over the Arctic, which is increasing in size and could affect Europe and North America.

Ultra-violet radiation is an important component of sunlight which, in small doses, is good for people, producing vitamin D. However, excessive exposure can be harmful.

Causes

The increased production and use of **chlorofluorocarbons** (CFCs) which:

- are light gases which rise slowly into the atmosphere
- can have a long 'life' of over 400 years
- attack and destroy ozone molecules
- are responsible for slowly eroding a hole in the ozone layer.

Chlorofluorocarbons (CFCs) are man-made chemicals used in foam packaging, aerosols and as a coolant in fridges and air-conditioning systems.

Fig 5.27 The ozone layer of the atmosphere

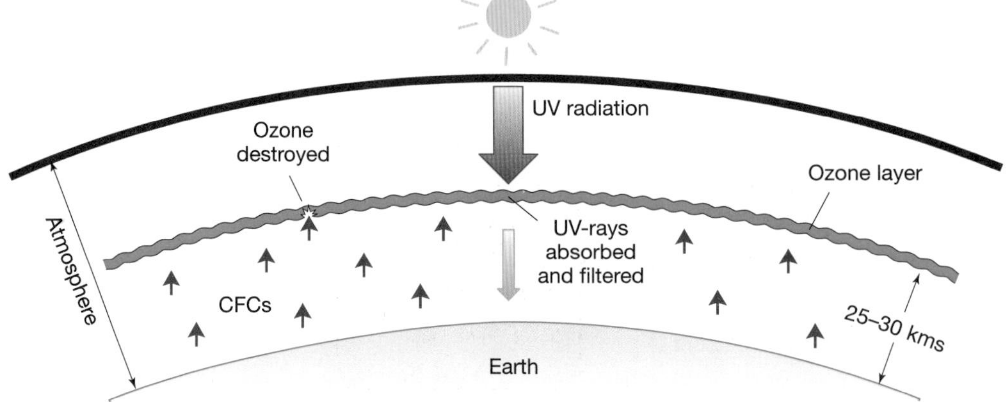

The Australian government and people treat this threat very seriously. 'Slap on the cream, slip on the hat.'

Consequences

In Australia and South America:

- skin cancer has increased; it is estimated that a 1% decrease in ozone causes a 5% increase in cancer
- eye cataracts have increased
- crops have been damaged by increased diseases
- marine food chains have been affected; micro-organisms that are important sources of food for fish are being destroyed.

Solutions

Sufficient CFCs are already in the atmosphere to cause erosion of the ozone layer until at least 2020.

- International Agreement in 1990 banned the use of CFCs by 2000:
  - most MEDCs have achieved this, some LEDCs require more time
  - new fridges no longer use CFCs, gases removed from old fridges, new 'ozone friendly' aerosols, biodegradable packaging has replaced foam
- Poster campaigns and the media have raised public awareness to use suntan creams and reduce exposure by always wearing sun-hats and long-sleeve shirts.
- Weather forecasts include information on UV intensities.

Photochemical smog

Smog is a thick, ground level fog caused when water droplets become polluted by chemicals and gases.

Affects cities in MEDCs and LEDCs.

Urban areas are experiencing an increase in the number of summer days with smog. Cities such as Calcutta, Mexico City, Athens and Los Angeles are badly affected.

Causes

Smog is more common:

- on clear, calm, hot summer days (higher UV radiation)
- where traffic is heavy and slow-moving (higher pollution levels)
- when cars are older with poorly maintained engines.

Fig 5.28 The formation of smog

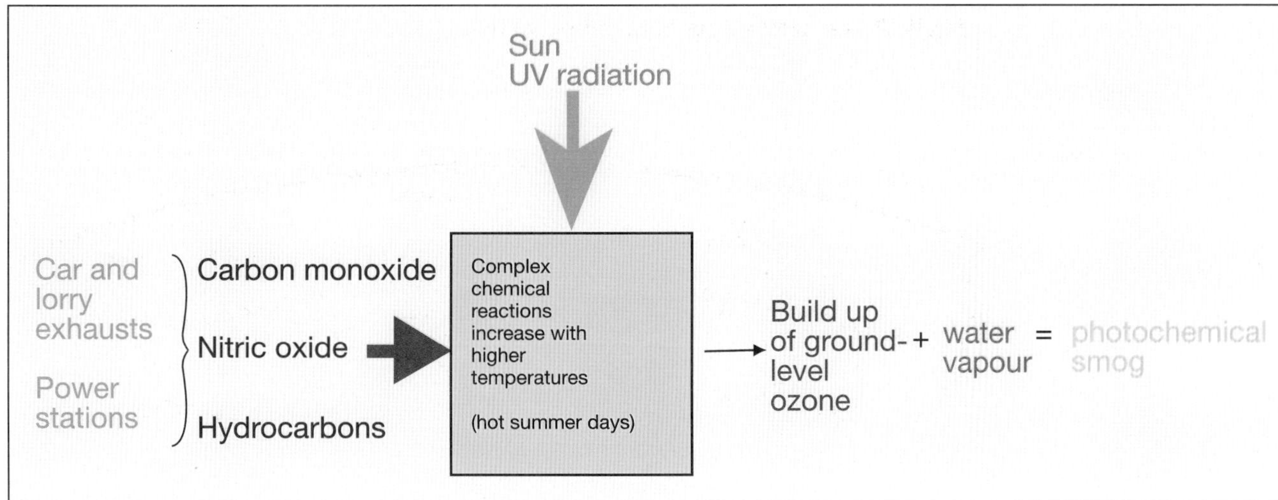

Consequences

- very unhealthy – increased hospital admissions
- causes irritation to the eyes and conjunctivitis
- causes breathing problems, increase in asthma attacks (can be fatal to older people)
- detrimental to plants and animals

Solutions

- restrict cars in city centres, (congestion charges, e.g. London) reduce lorry traffic
- improve and monitor engine efficiency
- improve electric-powered cars
- improve facilities for bicycles

Acid deposition

KEY POINT **Acid deposition is the fall-out from the atmosphere of acid compounds, either in rain as acid rain, or in very small particles as dry deposition.**

Acid rain was first recorded in Scandinavia during the 1950s. It is now recognised as affecting a large number of other countries including Canada, the USA, Germany, Scotland and Poland.

Acidity: pH is a measure of acidity or alkalinity.

Fig 5.29

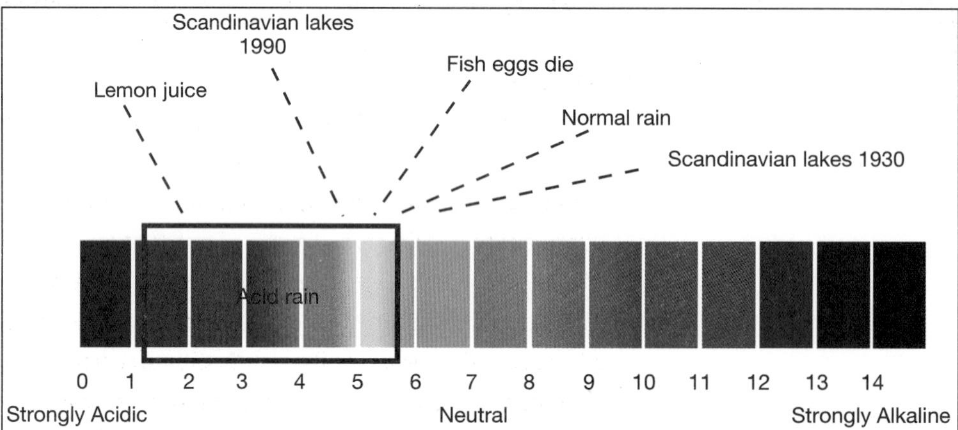

Causes

Fig 5.30 Causes of acid deposition

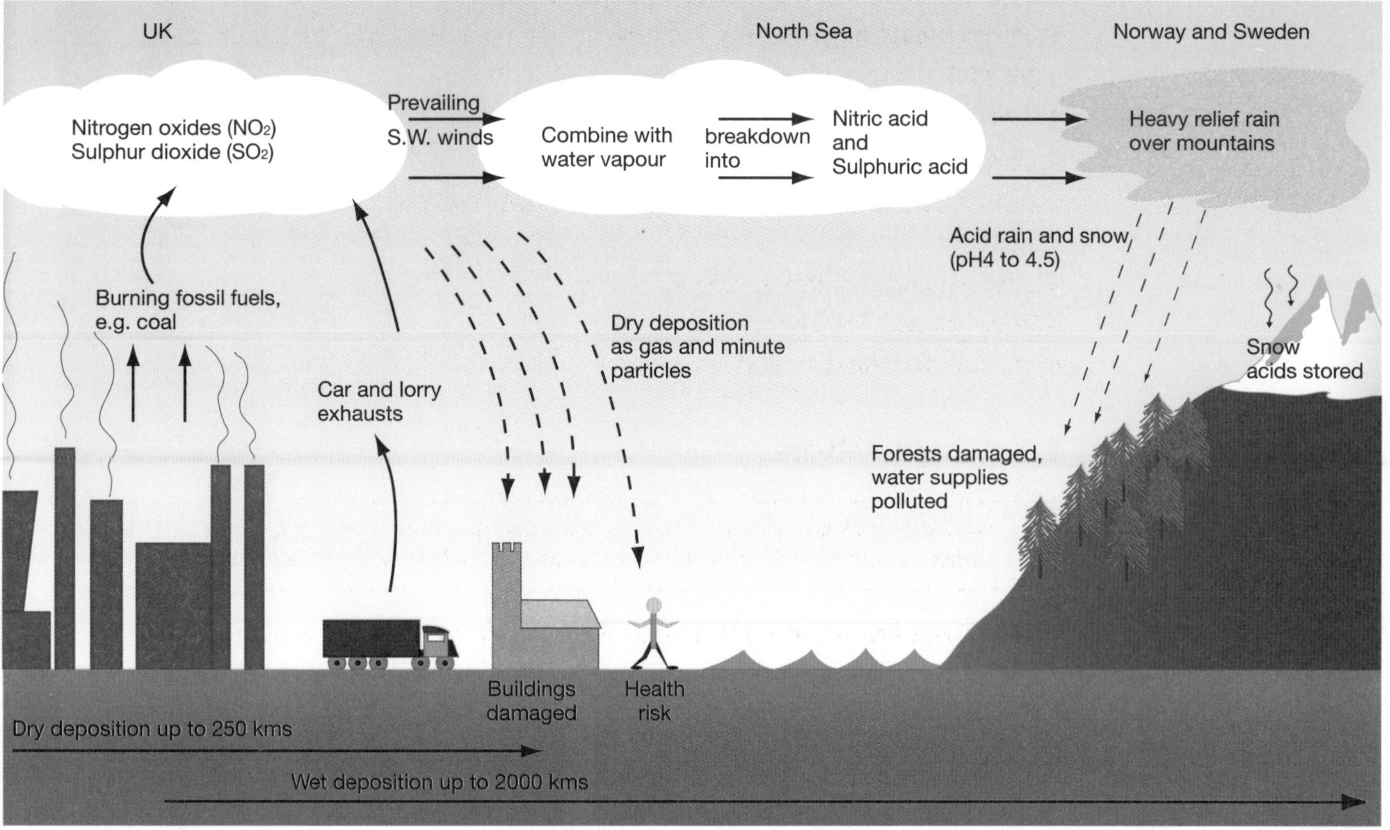

Consequences

- Acid rain releases nutrients in the soil which are carried away by percolating water (leaching); plants and trees suffer severe stress.
- Coniferous forests are dying; acid attacks the pine needles causing them to drop; greater risk from disease. Twenty-five per cent of the forests in southern Sweden, Germany and Poland have been severely damaged.
- Lakes and streams become so acid that they can no longer support any life at all, plants, fish or amphibians. (Meltwater from snow is very acid.)
- Lichens store chemicals which, when eaten by mice, rabbits and deer, contaminate the food chain (foxes, birds of prey).
- Acid soils restrict crop growth and reduce yields.
- Health hazard – domestic water supply is contaminated; acid water attacks copper pipes making water toxic.
- Stonework on buildings attacked (chemical weathering – particularly limestone and marble) e.g. Houses of Parliament, Taj Mahal in India.
- Dry deposition can affect health: respiratory problems and headaches.

Affects both MEDCs and LEDCs.

Solutions

- Increase height of power station chimneys – reduces local dry deposition but increases acid rain.
- Reduce sulphur dioxide emissions from power stations by:
  - burning low sulphur coal – expensive to import
  - fitting filters to chimneys – expensive, could increase price of electricity
  - change to gas – increasing use of natural gas in recent years, and decrease in coal, have reduced emissions by over 40% since 1988 in the UK.
- Reduce nitric oxide emissions from cars by fitting catalytic converters.
- Use alternative energy sources – hydro, solar, wind.
- International Agreements.
- Add lime (alkali) to lakes to restore pH – only a temporary solution.
- Increase public awareness – reduce energy demand and use of cars.

Solutions are mostly the responsibility of the countries producing the pollution.

1. What is the difference between weather and climate?
2. Which instrument is used to measure air pressure?
3. What is the name given to a line that joins places of equal rainfall?
4. Is air rising or sinking in an anticyclone?
5. Why is the south-west of the UK relatively warm in winter?
6. Which of the following areas of the UK receives the highest average precipitation per year:
   - the London area
   - the north-east
   - the Lake District?
7. What is the name given to the type of climate experienced in Uaupes?
8. Why is the layer of ozone found in the upper atmosphere beneficial to life on Earth?
9. Which of the following are causing global warming:
   - the hole in the ozone layer
   - the increase in greenhouse gases
   - the rise in sea level?
10. What is desertification?

1. Weather describes atmospheric conditions at a certain time and place; climate is the average weather of a place over many years. 2. Barometer 3. Isohyet 4. Sinking 5. Influence of the warm ocean current called the North Atlantic Drift 6. The Lake District 7. Tropical Equatorial 8. Protects us from exposure to too much ultra-violet radiation 9. The increase in greenhouse gases 10. The process caused by natural and human factors that turns land into desert

Chapter 6

Ecosystems

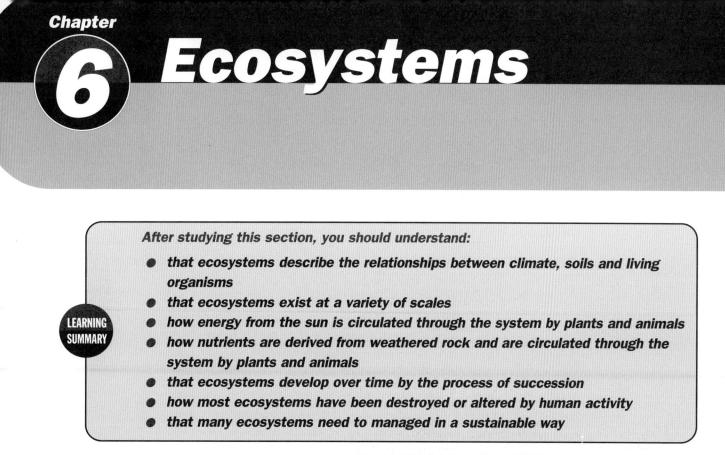

LEARNING SUMMARY

After studying this section, you should understand:

- ***that ecosystems describe the relationships between climate, soils and living organisms***
- ***that ecosystems exist at a variety of scales***
- ***how energy from the sun is circulated through the system by plants and animals***
- ***how nutrients are derived from weathered rock and are circulated through the system by plants and animals***
- ***that ecosystems develop over time by the process of succession***
- ***how most ecosystems have been destroyed or altered by human activity***
- ***that many ecosystems need to managed in a sustainable way***

The natural environment

The **natural environment** consists of the **rocks** and **soils**, **vegetation** and **animals**, and the **air** and **water** of the Earth; it is made up of different **ecosystems**.

An ecosystem describes the links between the living community of animals and plants and their habitat.

- The living community is the **biomass** (animals, plants, fungi, bacteria).
- They are linked to each other in a **food chain**.
- The **habitat** includes the non-living components such as rocks, soil, water and climate.
- Ecosystems exist at a variety of scales, from a rock pool on the seashore to **biomes**, which are large global systems, such as the tropical rainforest.
- Human activity has destroyed or altered parts of most ecosystems.

Energy cycle

Understanding the basic ideas of the energy cycle, nutrient cycle and succession will help you to explain why biomes are so different.

- The sun is the source of all energy in an ecosystem.
- Light energy from the sun is absorbed by green plants in a process called **photosynthesis**.
- Energy flows through the system as a food chain. Plants are eaten by animals, and animals eat each other (see Fig 6.10).
- The food chain consists of a series of stages known as trophic levels.
- Energy is stored in the system: in plant tissue, muscle and fat.
- Energy is lost from the system by transpiration, respiration (by plants and animals), consumption (when a worm is eaten by a bird) and decomposition (bacteria and fungi break down dead plants and animals).
- Human activity can interfere with the food chain, e.g. deforestation, keeping large flocks of sheep, hunting large carnivores, introducing new species.

Fig 6.1 An ecosystem

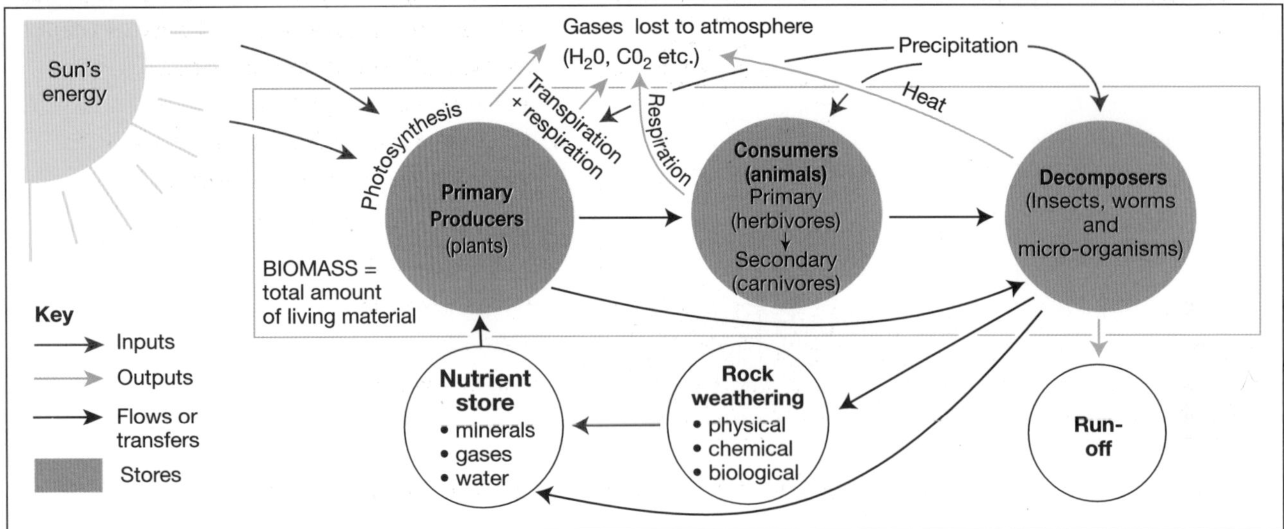

Fig 6.2 Trophic levels of an ecosystem

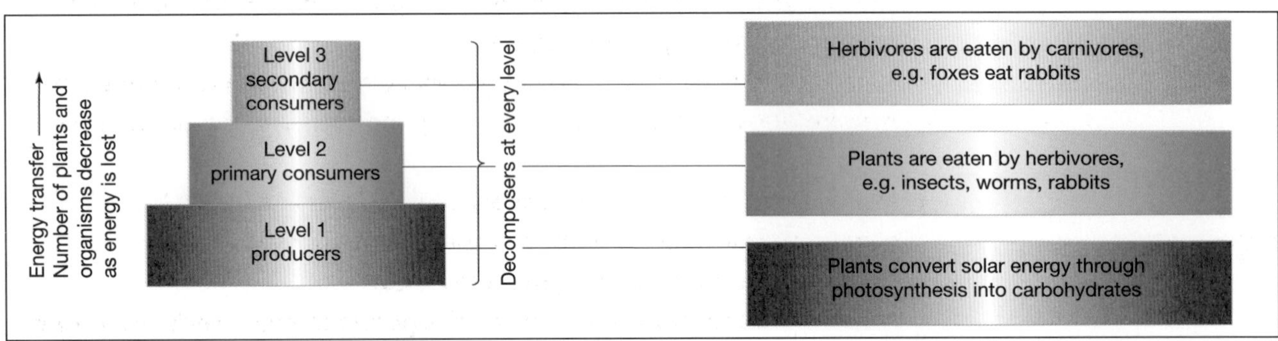

Nutrient cycle

- Weathered rock releases nutrients such as calcium, carbon and nitrogen into the soil.
- Plants and animals circulate nutrients through the system.
- The cycle occurs quickly in hot, wet ecosystems, e.g. tropical rainforest, and more slowly in cold systems, e.g. coniferous forest.
- Nutrients are lost by percolation and run-off into rivers.
- Human activity can affect the cycle by deforestation – loss of plant material and increased run-off.

Fig 6.3 The nutrient cycle

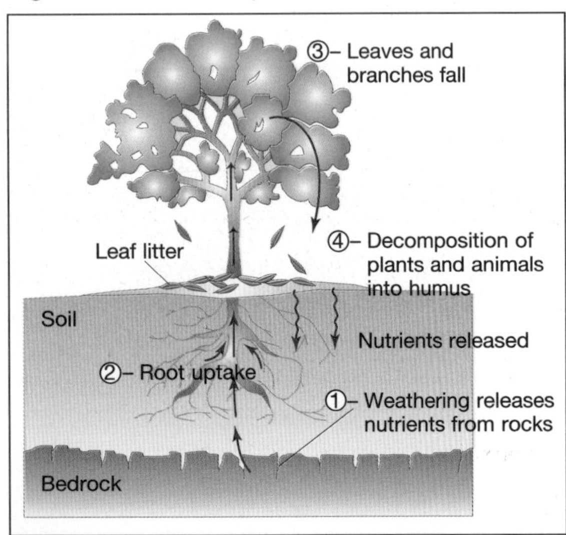

Succession

Ecosystems develop over time by the process of succession.

- The first plants to colonise an area are called pioneer species.
- Pioneer plants are adapted to harsh conditions. They are hardy and require only a low nutrient supply.
- Rock weathering and the decomposition of plants increase the supply of nutrients, allowing new plants to germinate, e.g. grasses. Insect and animal life increases.
- Soils become deeper, more nutrients – larger and more varied species, e.g. shrubs.

Fig 6.4 Typical succession to climax vegetation

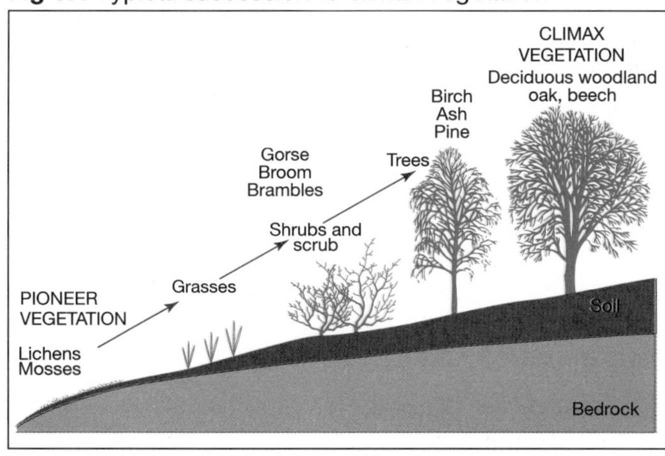

- Given time, a dominant species, such as oak or beech trees, take over and smaller plants live beneath the dominant trees. The succession is completed.
- Plants, insects, birds and animals exist in complex food chains and webs.
- There are four types of succession:
  - on bare rock
  - in freshwater
  - on sand
  - in salt water.

A small-scale ecosystem – sand dunes

- Dunes form along coasts where there is a large supply of sand from the beach and strong prevailing winds blowing in from the sea. Dunes are common on sand spits (see Fig 3.9).
- Sand is blown inland forming low embryo dunes which are colonised by marram grass. Further sand is trapped by the grass and the dunes grow higher.
- Marram grass is **adapted** to the dry and windy environment by:
  - growing fast so as not to be smothered by fresh sand
  - having long roots to reach the water table
  - having strong leaves that can stand exposure to high winds and prevent excessive moisture loss.
- Marram grass helps to stabilise the dunes – decomposition provides humus for soil development.
- Older dunes are more stable and have a wider variety of plants covering the ground. Shrubby plants develop, e.g. sea buckthorn, hawthorn, gorse, which are colonised by insects, birds, rabbits, birds of prey and foxes. Pine trees and eventually the dominant species of oak and ash complete the succession.

The plants and animals interact with the wind and sand, and the succession develops.

Fig 6.5 Succession to climax vegetation on sand dunes

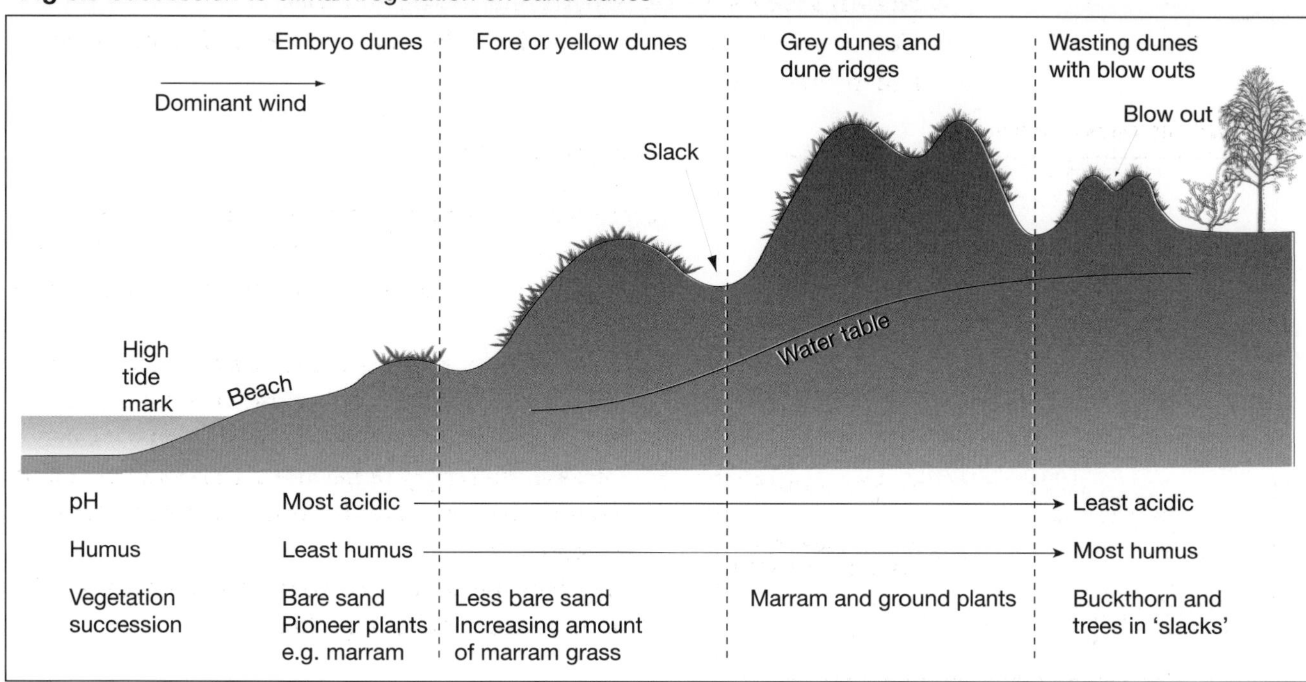

Dune management

- Dunes are important coastal defences and need to be conserved.
- Dunes are fragile systems and can quickly be damaged by:
  - trampling, sliding down dune faces, horse riding and fires
  - over-population by rabbits
  - military use.
- Damage to dunes leads to:
  - vegetation being destroyed
  - bare sand exposed
  - wind removes sand, leaving 'blow-outs', exposing root systems which kills marram grass
  - greater exposure of bare sand – more wind erosion.
- Management techniques include:
  - wooden boardwalks and fencing along popular routes through dunes
  - information/interpretative boards to raise public awareness
  - wardens and guides
  - car parking charges to reduce demand
  - culling and controlling rabbit populations.

Sand dunes at Gibraltar Point, Skegness (Fig 3.12) and Studland (Fig 3.14) are carefully managed to allow public access and sustain the dune habitat.

Biomes

AQA A AQA B

KEY POINT **Biomes are large ecosystems at the global scale where the climate, vegetation and soils are broadly the same.**

- Human activity has destroyed or changed parts of all biomes.
- There are eight major biomes:
  - Tundra
  - Mediterranean
  - Coniferous forest (taiga)
  - Desert
  - Temperate deciduous woodland
  - Savanna grassland
  - Temperate grasslands
  - Tropical rainforest

Fig 6.6 Four of the major biomes

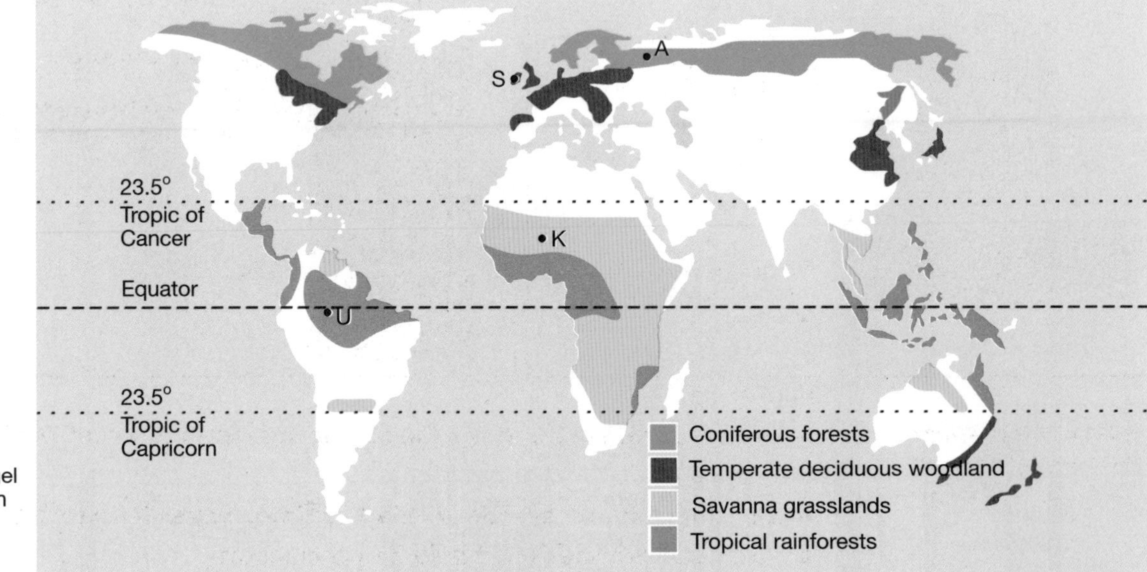

Key
A = Archangel
S = Shannon
K = Kano
U = Uaupes

Fig 6.7A Climate graph for Archangel ($64\frac{1}{2}$°N)

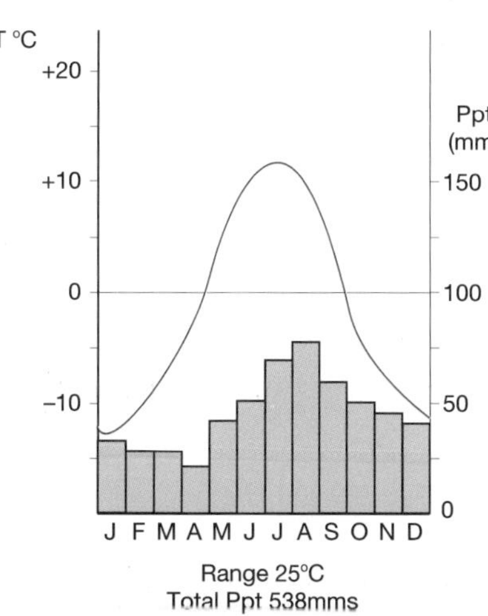

Note how the vegetation is adapted to the cold climate.

Coniferous forest

Location (see Fig 6.6):

- found only in the northern hemisphere in a wide band between 50°N and the Arctic Circle; found in USA, Canada, northern Europe and Siberia

Climate:

- long, cold winters (average temperature -15°C) with extreme temperatures (-30°C) and frosts. Short summers with long days – short **growing season** (three months)
- low precipitation (500mm) – cold air unable to hold much water vapour. Summer maximum, winter snowfall melts only in spring

Vegetation and animals:

- spruce, pine and fir: large areas of single species, few other plants owing to thick layer of pine needles and lack of light all year from tree cover
- **evergreen**, keeping leaves (needles) all year
- small needle-shaped leaves, waxy, to reduce transpiration
- conical tree shape
- seeds protected in cones
- voles, hares, reindeer, wolves, bears – few birds except in summer

Soils:

- **podsols**: shallow acidic soils; nutrients and minerals washed downwards (**leaching**) by percolating water in summer
- pine needles slow to decompose (low temperatures) – thin humus layer
- clear layers (**horizons**), few worms

Fig 6.7B A coniferous tree

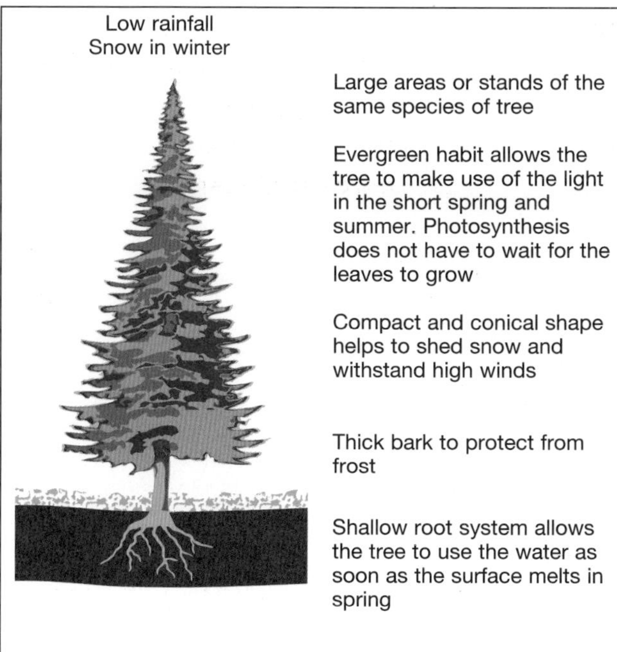

Fig 6.7C Podsol soil profile

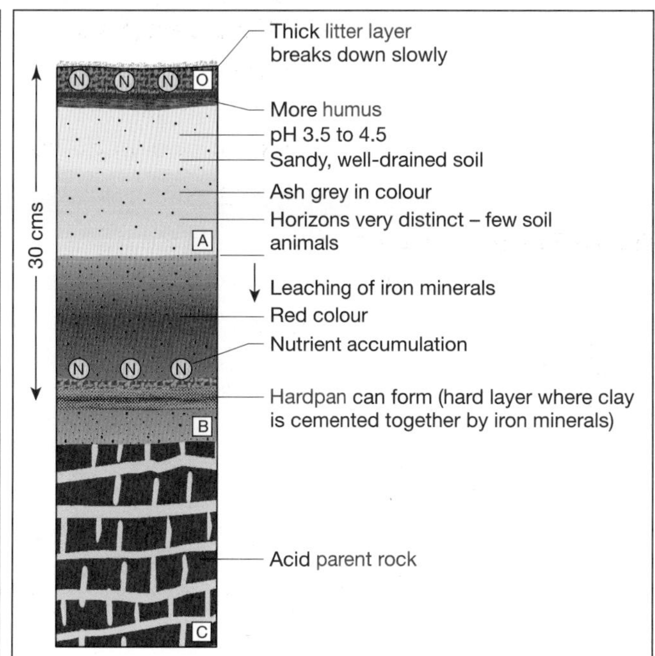

Research sustainable forestry strategies in Sweden.

Human use:

- very productive because of large **biomass**; most forests commercially managed
- tall, straight trunks provide excellent softwood planks for building industry
- **softwood** provides wood pulp for paper industry

Fig 6.8A Climate graph for Shannon, Eire (52½°N)

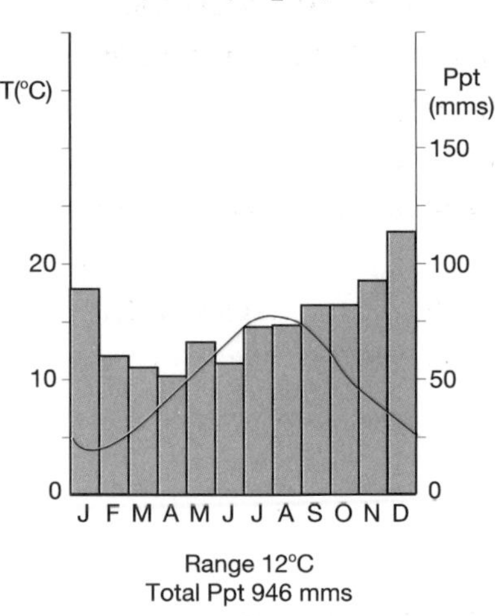

Temperate deciduous woodland

Location (see Fig 6.6):

- mostly on the east coast of continents, between 40°N to 60°N, but also includes north-west Europe, including the UK and Eire (see page 65, Fig 5.15).
- includes eastern USA, much of China, Japan and New Zealand

Climate:

- cool summers – growing season is seven months; mild winters
- precipitation 500–1000mm/year, falling throughout the year

Vegetation and animals:

- **deciduous** trees (oak, beech, ash): shed leaves in winter (reduces transpiration when colder temperatures reduce effectiveness of **photosynthesis**)
- relatively few species compared with tropical rainforest
- oak represents **climax vegetation**
- there are some layers beneath the upper canopy as light can penetrate, particularly in spring (shrub layer, field layer)
- thick leaf litter, but many decomposers cause rapid recycling of nutrients
- deer, foxes, squirrels, badgers, many bird species

Fig 6.8B Deciduous woodland

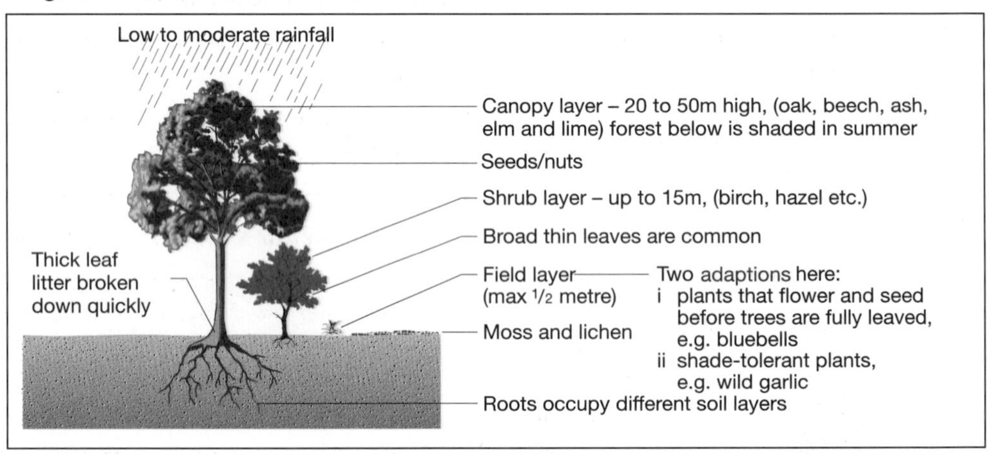

Fig 6.8C Brown earth soil profile

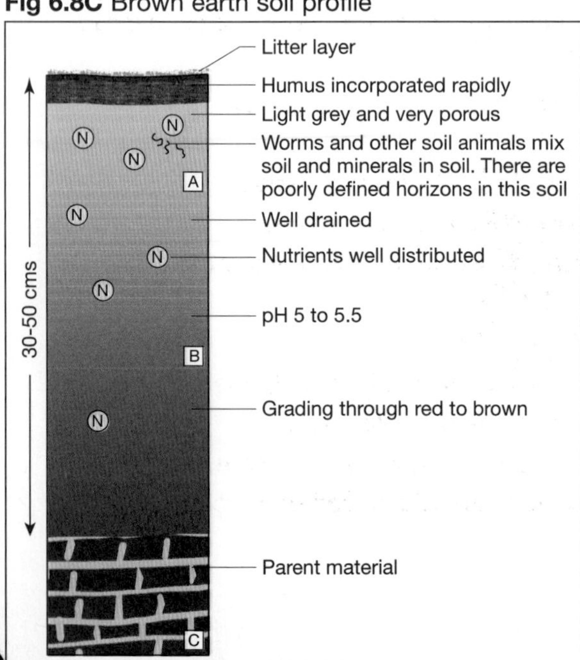

Soils:

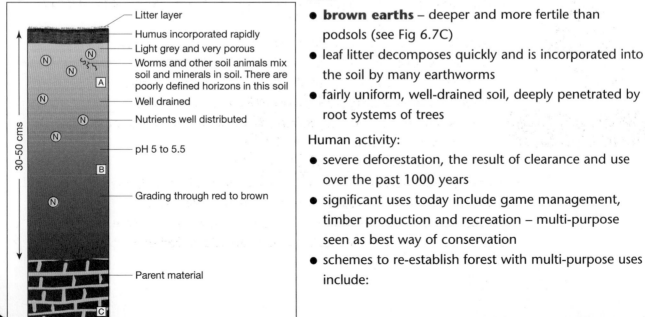

- **brown earths** – deeper and more fertile than podsols (see Fig 6.7C)
- leaf litter decomposes quickly and is incorporated into the soil by many earthworms
- fairly uniform, well-drained soil, deeply penetrated by root systems of trees

Human activity:

- severe deforestation, the result of clearance and use over the past 1000 years
- significant uses today include game management, timber production and recreation – multi-purpose seen as best way of conservation
- schemes to re-establish forest with multi-purpose uses include:

Carry out research to find how these schemes are progressing.

- **Community Forest Organisation** – since 1991, 3500 hectares of native trees have been planted in twelve community forests in the UK, mostly on brownfield sites on the edges of cities, e.g. Great North forest (Newcastle), Mersey forest (Liverpool – Manchester)
- **National Forest** – designated in 1990 with aims to:
  - plant a third of the Midlands (Staffordshire, Derbyshire, Leicestershire) with the new woodland
  - use **brownfield sites**, e.g. former mine tips, for commercial forestry, leisure pursuits, e.g. cycling, fishing, walking, and wildlife reserves.

Fig 6.9A Climate graph for Kano, Nigeria (12°N)

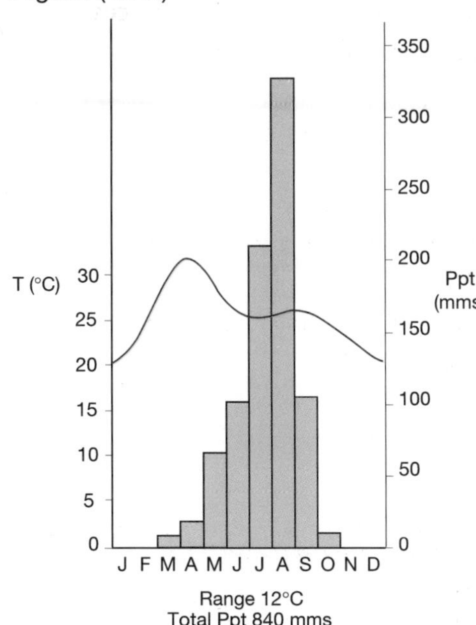

Savanna grasslands

Location (see Fig 6.6):

- found between 25°N and 25°S of the Equator, mostly covers the central parts of continents
- includes central Africa (northern Nigeria, Kenya, Zambia,), north Australia and central South America (Venezuela, Brazil)

Climate:

- temperatures are high throughout the year (average 25°C) with a short cooler season
- **wet and dry seasons** – summer wet season when sun gives maximum heat (see Fig 5.12) – 80% of rainfall occurs in four months as heavy convectional storms (thunder and lightning)
- little rain in dry season with dry winds blowing from the Sahara Desert (see Fig 5.23)
- wet season may be becoming unreliable due to climatic change leading to **desertification** (see Chapter 5, page 72)

KEY POINT **Savanna is transitional between tropical rainforest (heavy rain all year) and desert (low unreliable rainfall)**

Vegetation and animals:

- mixture of trees and grasses, with grasses dominating towards the desert
- deciduous trees, shedding leaves in dry season to limit transpiration and loss of water, e.g. baobab tree
- trees have adapted to drought by having:
  - small, waxy or thorn-like leaves
  - large trunks incorporate fleshy cells capable of storing water
  - long roots, extending down to water table, e.g. palm trees
  - thick, fire-resistant bark (fires are common as a result of lightning during convectional storms)
- grasses grow in tufts, with thick strong blades
- grasses grow quickly with the first rains, up to 5m in height, and die back in the dry season; seeds remain dormant until rain falls again
- supports many species of **herbivore** such as zebra and antelope (Fig 6.10)
- supports **carnivores** such as lion and hyena which feed on herbivores

Fig 6.9B Savanna grassland section

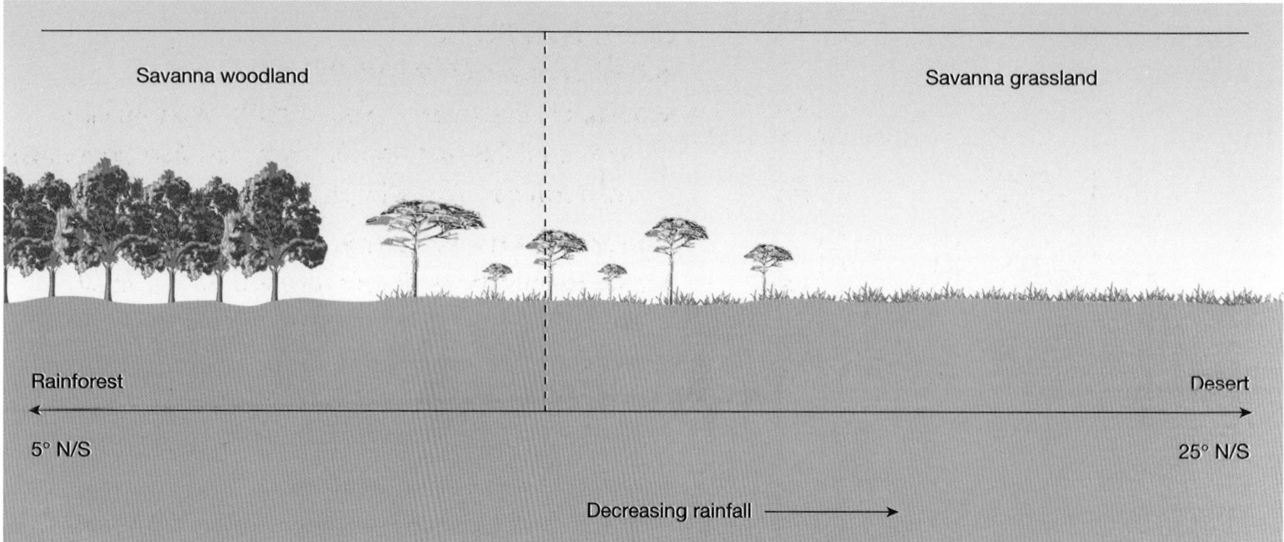

Fig 6.9C Savanna soil profile

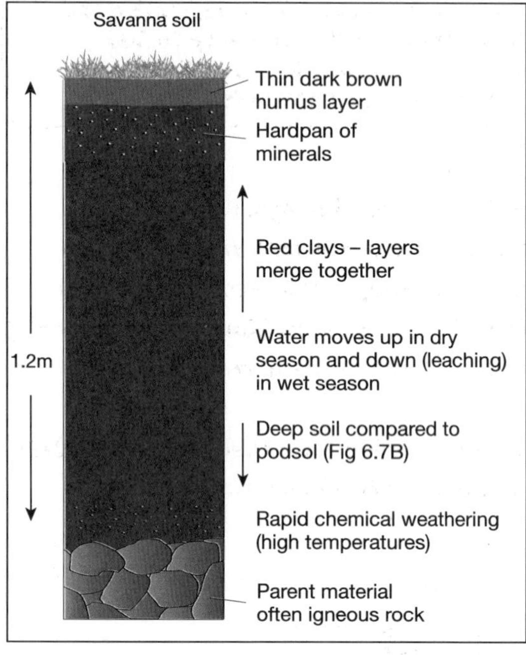

Soil:

- soils are red-coloured clays, rich in iron and aluminium oxides
- dead leaves and grass rapidly decay in high temperatures but give little humus or nutrients (low fertility)
- hard layer ('pan') develops just below surface, restricting penetration of water and roots – makes ploughing difficult and, if vegetation is removed, the soils are easily eroded by the wind.

Human activity:

- most suitable for nomadic pastoralism – extensive, low density cattle and sheep grazing, following the wet season (fresh grasses), north and south
- crops can be grown, particularly where land is irrigated – maize, millet and tobacco
- increased population and decline in nomadism (settling around water holes) lead to overgrazing – removal of vegetation by increased wind erosion and desertification

Fig 6.10 Simple food web in savanna grassland

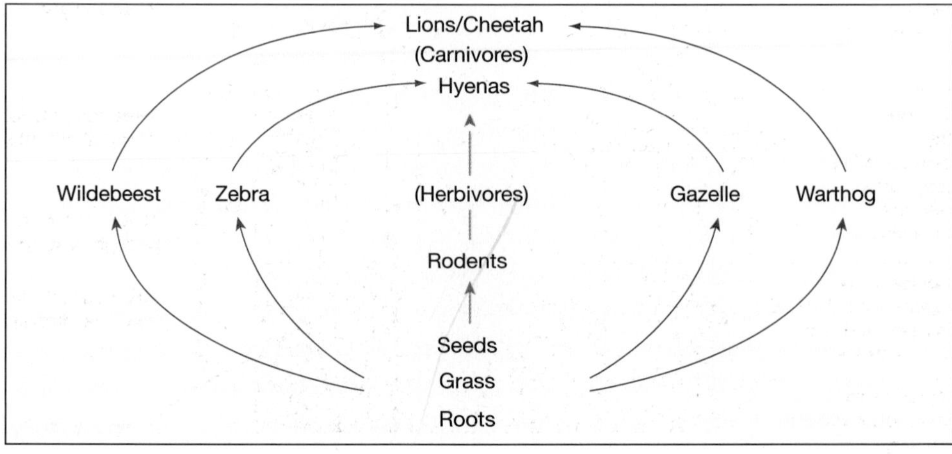

Fig 6.11A Climate graph for Uaupes, Brazil (1°S)

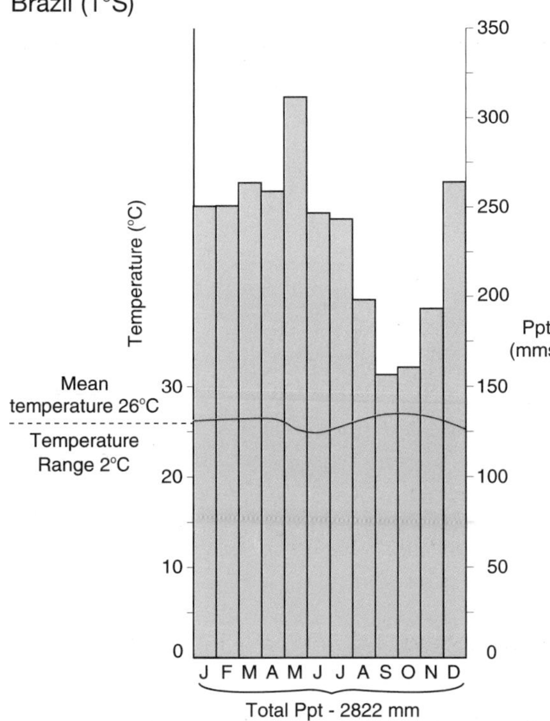

Tropical rainforests

Location (see Fig 6.6):

- found between 5°N and 5°S of the Equator
- areas include Amazon Basin (Brazil), West Africa (southern Nigeria, Congo), South East Asia (Indonesia) and north-east Australia (state of Queensland)

Climate (see Chapter 5, page 67):

- high temperatures all year (average 26°C); small annual temperature range (2°C); 12 month continuous growing season
- annual rainfall very high (over 2000mm), falling throughout year (some areas slightly away from Equator have a short 'drier' season)
- high humidity and rising air gives daily convectional storms (in the afternoon)

The biome with the most diverse vegetation and animal life: Amazon rainforest has up to 300 species of tree per square km – mostly hardwoods, e.g. ebony, teak, mahogany

Vegetation and animals:

- **deciduous** trees, growing and shedding leaves throughout year, gives evergreen appearance; flowers, fruits and seeds available all year from different species of tree – huge continuous supply of leaves to forest floor
- distinct layers of vegetation with the tallest trees up to 50m tall (**emergents**)
- trees have shallow roots but develop **buttress roots** above ground to support their great height
- leaves are thick and waxy to reduce transpiration if exposed to the sun; they have **'drip tips'** in order to shed the heavy rainfall efficiently

Fig 6.11B Cross-section through a tropical rainforest

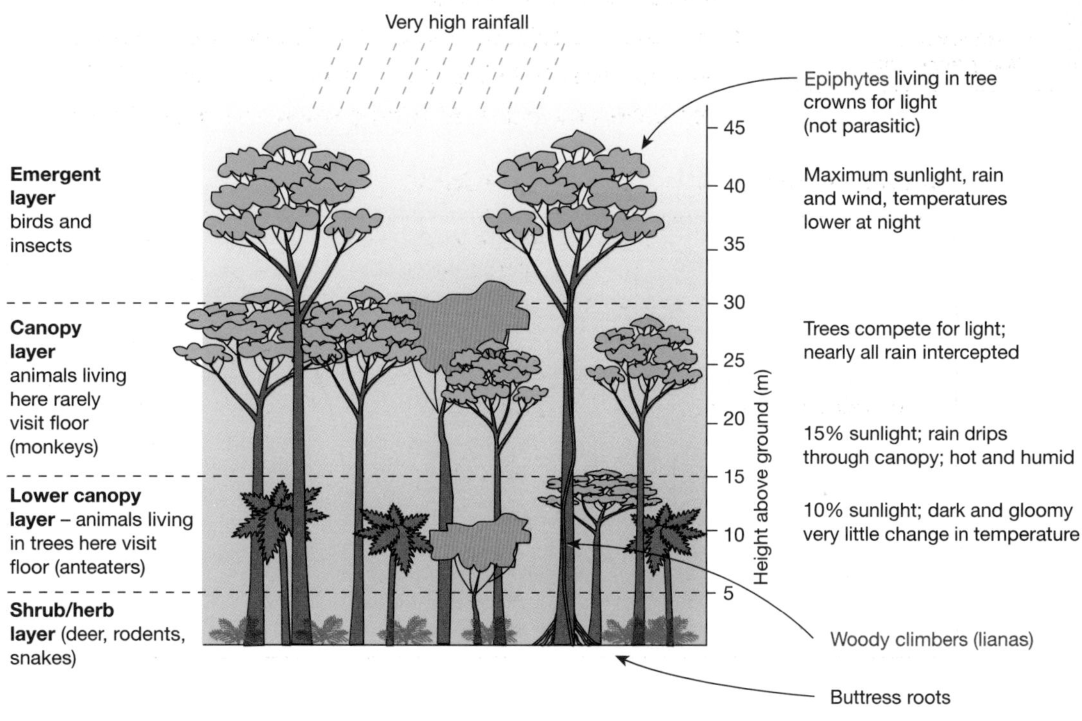

Fig 6.11C Tropical soil profile

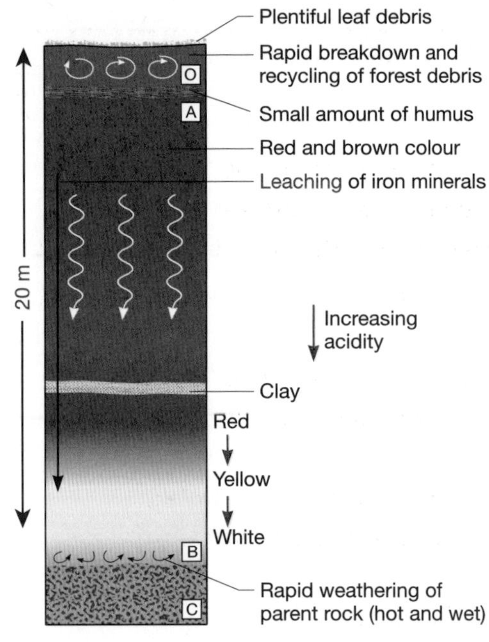

- **epiphytes** grow on trunks and branches, but do not feed on the trees' nutrients
- at ground level the forest is dark, little sunlight penetrates the dense canopy, undergrowth is therefore limited – dense growth if trees are cut down
- huge variety of animals occupy different levels of the forest, (rich all year food supply) – rivers contain many species of fish and reptiles

Soils:

- red clay soils rich in iron and aluminium oxides
- large numbers of decomposers
- high temperatures and humidity cause rapid chemical weathering:
  - rapid breakdown of leaf litter, fast recycling of nutrients through dense root systems; soils keep few nutrients and are infertile; nutrients are stored in the biomass
  - rapid weathering of parent rock gives very deep soils – up to **20 metres**
- soils have a weak structure and erode easily if exposed to heavy rain, e.g. deforestation

Human activity:

- **shifting cultivation** (slash and burn) – traditional form of agriculture followed by local (**indigenous**) farmers (see also Chapter 12)
- farming is supplemented by hunting, fishing and gathering (different fruit available all year round)
- extensive areas of forest are required to support a small number of families
- in many areas the system is under pressure from:
  - population growth: cleared areas are not given sufficient time to recover before being used again (**fallow period** is reduced) – soil erosion more likely and yields decline
  - large areas of forest are being cut down for logging and other new developments, reducing the forest available for traditional farmers

Fig 6.12 Flow diagram of shifting cultivation in the rainforest

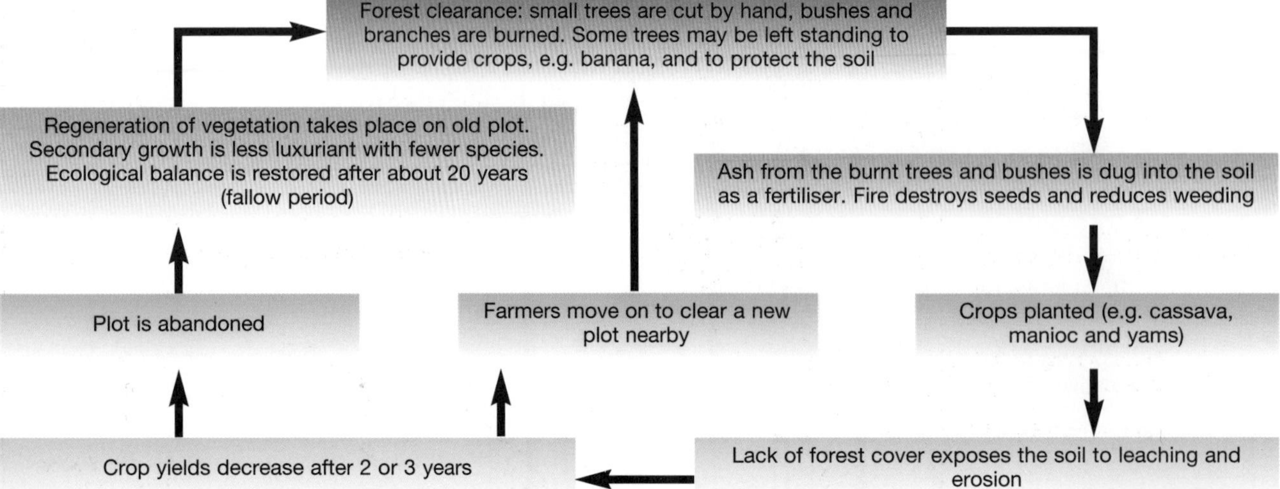

Deforestation of tropical rainforests

Deforestation is the cutting down and removal of areas of forest, in any part of the globe.

- Fifty per cent of tropical rainforest has been destroyed in the last 100 years.
- The rate of destruction has been increasing rapidly in the last twenty years.
- Some scientists estimate that if the current rate of clearance continues, most forests will have disappeared in the next ten years.

Causes of deforestation

Link causes to examples in Brazil or another LEDC you have studied.

- Extension of commercial agriculture – forest is cleared for cattle ranches.
- Timber extraction – hardwood for MEDCs, e.g. Japan and UK.
- Mining operations.
- Developing electricity supplies – HEP projects (reliable high rainfall).
- Road building – new roads allow easier access in and out the forest.
- Population growth:
  - within forest communities (shifting cultivators)
  - government resettlement schemes of urban poor, alongside the new roads
  - exploitative recreation and tourism – new hotels and leisure complexes.

Fig 6.13 Causes of deforestation in the Brazilian rainforest

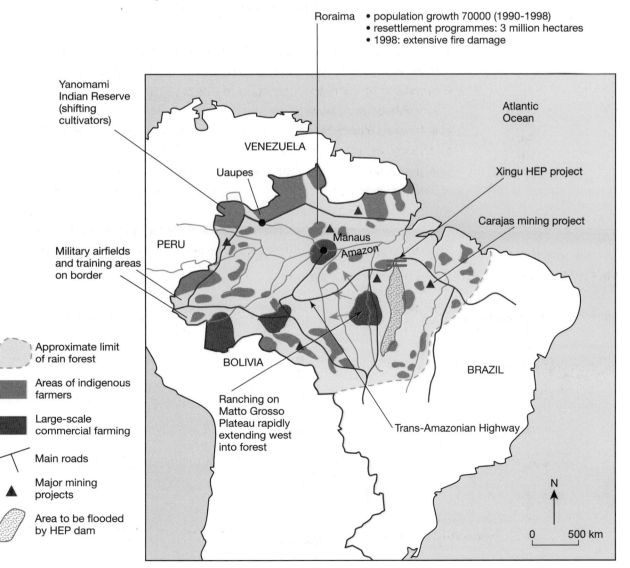

Roraima
• population growth 70000 (1990-1998)
• resettlement programmes: 3 million hectares
• 1998: extensive fire damage

Yanomami Indian Reserve (shifting cultivators)
Atlantic Ocean
VENEZUELA
Uaupes
Xingu HEP project
Carajas mining project
Manaus
Amazon
PERU
Military airfields and training areas on border
BOLIVIA
BRAZIL
Ranching on Matto Grosso Plateau rapidly extending west into forest
Trans-Amazonian Highway
N
0 500 km

Approximate limit of rain forest
Areas of indigenous farmers
Large-scale commercial farming
Main roads
Major mining projects
Area to be flooded by HEP dam

Benefits of deforestation

- Most rainforest is located in LEDCs, e.g. Brazil, Indonesia:
  - money earned from export of timber and ores will increase the GDP and lead to economic and social development and improve quality of life (schools, hospitals)
  - LEDCs require income to pay off loans from MEDCs.
- Jobs are provided by mining, logging and tourism.
- Resettlement provides better quality of life for people living in urban 'shanty towns'.

Problems of deforestation

- Nutrients are stored in the biomass. If trees are cut down, an infertile soil is left. New farmers will only obtain satisfactory yields for three or four years. After that, they will need to cut down more forest or abandon the farm.
- Exposed soils are quickly eroded by heavy rainfall. Clay soil is washed into rivers which causes deposition and flooding, making navigation difficult, ruining fishing and polluting water supplies.
- Rapid run-off of heavy rainfall on exposed areas causes flash floods and bank erosion.
- Indigenous people are losing their livelihood, homes and culture. They fall victim to diseases introduced by settlers which cause high death rates.
- Vaccines for diseases such as malaria have been developed from rainforest plants. Deforestation may prevent the discovery of new plants with possible medical value.
- Deforestation may be affecting climate change:
  - trees absorb CO_2 – loss of trees will increase the concentrations of CO_2 (greenhouse gas) in the atmosphere – a possible cause of global warming
  - burning trees releases more CO_2 into the atmosphere

Note the conflict between the need for LEDCs to develop and improve the quality of life of their people and global environmental considerations.

MEDCs suggest that LEDCs stop deforestation; accusing them of increasing global warming. But MEDCs demand most timber and are the largest importers. MEDCs burn 80% of the world's fossil fuels, which are the principle cause of global warming.

Solutions for the Amazon rainforest

Proposals include sustainable conservation schemes such as:

How are these plans sustainable? What are the advantages and disadvantages of these proposals for different groups such as indigenous farmers, loggers, the government, Friends of the Earth?

I Agroforestry

- Technique imitates the protection 'layers' of the natural forest, which protects the soils from erosion.
- Maintains the humus layer in the soil and improves fertility.
- Retains the surrounding forest for hunting and gathering.
- Provides annual crop-for-cash sales and longer term tree resources.
- Supports a large population that can be self-sufficient.
- Discourages non-indigenous 'new settlers'.

Fig 6.14 Agroforestry

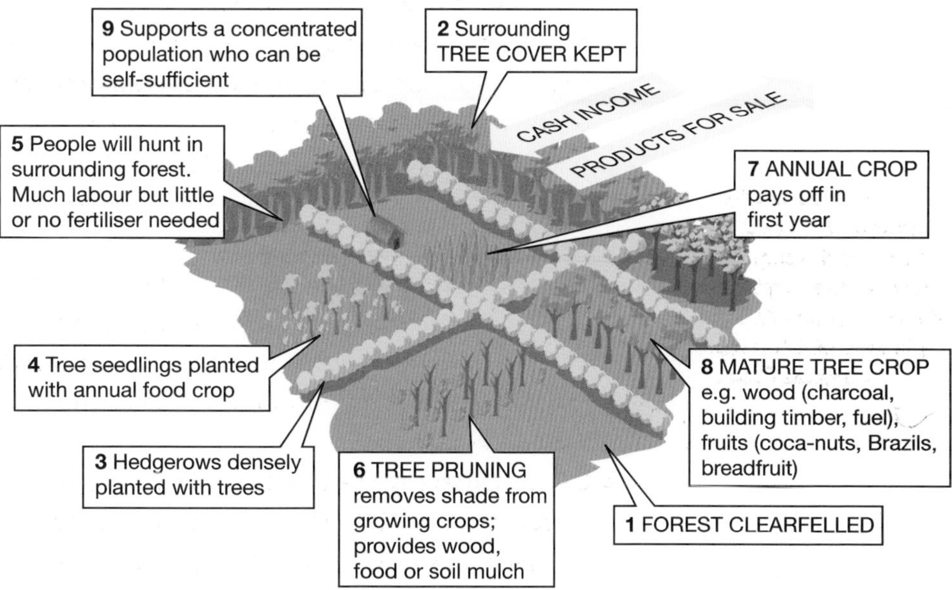
9 Supports a concentrated population who can be self-sufficient
2 Surrounding TREE COVER KEPT
CASH INCOME
PRODUCTS FOR SALE
5 People will hunt in surrounding forest. Much labour but little or no fertiliser needed
7 ANNUAL CROP pays off in first year
4 Tree seedlings planted with annual food crop
8 MATURE TREE CROP e.g. wood (charcoal, building timber, fuel), fruits (coca-nuts, Brazils, breadfruit)
3 Hedgerows densely planted with trees
6 TREE PRUNING removes shade from growing crops; provides wood, food or soil mulch
1 FOREST CLEARFELLED

II Biosphere reserve

- Establishes three 'zones', which have designated non-intrusive uses.
- Retains a large area of 'natural' forest for research.
- Maintains a secure zone for indigenous shifting cultivators.
- Provides an income from harvesting tree crops.
- Minimum disturbance of the forest.

Fig 6.15 Biosphere reserve

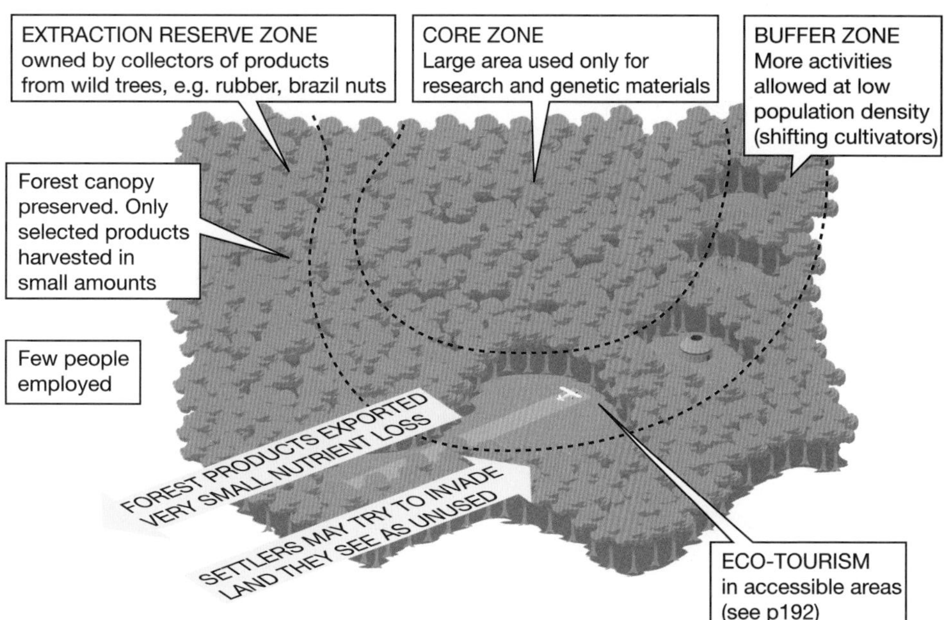
EXTRACTION RESERVE ZONE owned by collectors of products from wild trees, e.g. rubber, brazil nuts
CORE ZONE Large area used only for research and genetic materials
BUFFER ZONE More activities allowed at low population density (shifting cultivators)
Forest canopy preserved. Only selected products harvested in small amounts
Few people employed
FOREST PRODUCTS EXPORTED VERY SMALL NUTRIENT LOSS
SETTLERS MAY TRY TO INVADE LAND THEY SEE AS UNUSED
ECO-TOURISM in accessible areas (see p192)

- Environmental groups such as the **World Wide Fund for Nature** and **Friends of the Earth** campaign to make the public aware and to persuade industry and governments to respond to the issues.
- **Forest Stewardship Council** – links timber producers to customers and gives assurances that timber and timber products are only from areas of sustainable logging.
- MEDCs contribute to share the 'cost' of conserving the rainforest, and ensure that the living standards of all people in LEDCs improve.

PROGRESS CHECK

1. What is an ecosystem?
2. By what process do green plants absorb light energy from the sun?
3. In a food chain, what is the name given to herbivores such as rabbits?
4. What is the principal source of nutrients in an ecosystem?
5. What name is given to plants which first colonise an area?
6. What name is given to global-scale ecosystems such as tropical rainforests?
7. What type of soil is found in coniferous forests?
8. What are the dominant trees in a temperate deciduous forest?
9. What types of vegetation are found in the savanna?
10. Where are tropical rainforests found?
11. Draw a climate graph using the following data. Use sensible scales and label the three axes carefully – see Fig 6.11A.

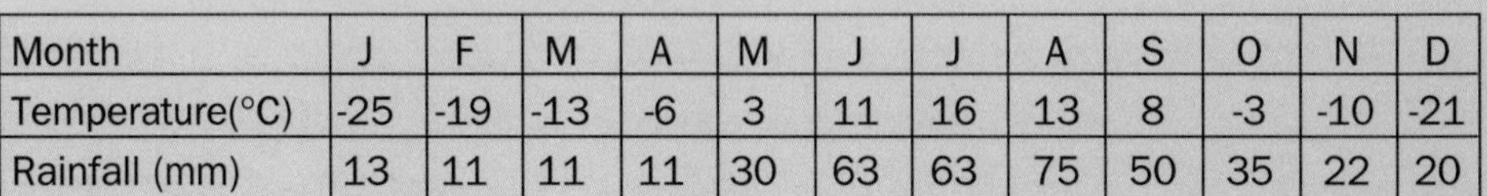

Month	J	F	M	A	M	J	J	A	S	O	N	D
Temperature(°C)	-25	-19	-13	-6	3	11	16	13	8	-3	-10	-21
Rainfall (mm)	13	11	11	11	30	63	63	75	50	35	22	20

12. What is the temperature range (°C)?
13. What is the total annual rainfall (mm)
14. In which biome would you be most likely to find this place? Give clear reasons using the data given.

1. Relationships between climate, soils and living organisms in an area 2. Photosynthesis 3. Primary consumers 4. Weathered rock 5. Pioneer species 6. Biomes 7. Podsol 8. Oak, ash, beech, elm 9. Drought-resistant trees and grasses 10. 5°N to 5°S of the Equator 12. Range: 41°C 13. Total rainfall: 404mm. 14. Coniferous forest (see page 85) (Data is for SURGUT, a city in central Russian Federation ($61\frac{1}{2}$°N). Locate with an atlas.)

Chapter 7 Tectonic activity

After studying this section you should understand:

- ***that the crust of the Earth is divided into plates, which can be divided into continental and oceanic plates***
- ***that tectonic activity results in earthquakes, volcanoes and fold mountains***
- ***that most tectonic activity is associated with the plate margins***
- ***that plate margins can be divided into three main types depending on how the plates are moving in relation to each other***
- ***that each type of plate margin is linked to different combinations of tectonic activity***
- ***the formation of fold mountains and their associated human activity***
- ***the main characteristics of earthquakes and volcanoes***
- ***the hazards and benefits of volcanic activity***
- ***the impact of earthquakes and volcanic activity in an LEDC and an MEDC***
- ***the management issues (such as prediction and preparedness) associated with reducing the impact of earthquake and volcanic activity***

Plate tectonics

To understand the location, distribution and causes of volcanoes and earthquakes, it is necessary to study **plate tectonics**.

Plate tectonics is the theory explaining the movement, formation and destruction of the plates which make up the Earth's crust.

The crust

The **crust** forms the **outer surface** of the Earth. The crust:

- is cool enough to behave as a more or less rigid shell
- is up to 90km thick (compared with a total radius of the Earth of 6400km)
- floats on hotter, semi-molten rock (the **mantle**) below
- is sub-divided into plates:
  - seven large plates, e.g. Pacific plate, Eurasian plate
  - twelve smaller plates, e.g. Philippine plate, Juan de Fuca plate

Plates

There are two types of plate:

- **oceanic**
- **continental**

Fig 7.1 The internal structure of the Earth

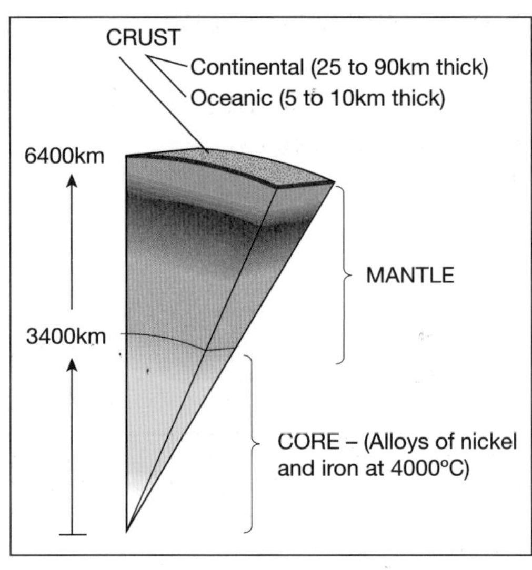

Fig 7.2 Principal crustal plates and plate margins

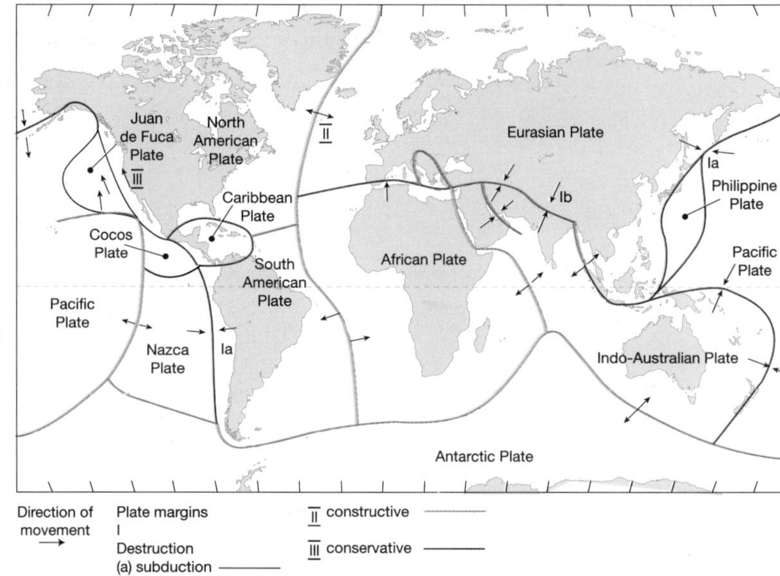

200 million years ago all the land masses of the Earth formed one continent known as Pangaea. Note how South America closely fits the shape of the west coast of Africa.

Oceanic plates are:

- composed of basaltic rock
- 5–10km thick
- denser than continental plates
- frequently destroyed beneath the continental plates
- younger than continental plates.

Continental plates are:

- mainly composed of granite
- 25–90km thick
- less dense and therefore can 'float' on the top of oceanic plates.

The plates are slowly moved by convection currents in the mantle.

Fig 7.3 How crustal plates move

Plate margins

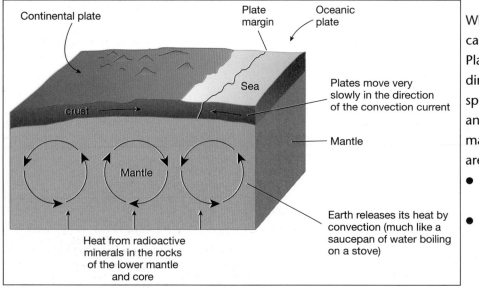

Where two plates meet is called a **plate margin**. Plates move in different directions and at different speeds, which causes stress and friction along the plate margins. Stress and friction are mostly responsible for:

- earthquakes and volcanic activity
- the creation of **fold mountains**, e.g. the Himalayas.

It is important to understand the difference between these margins. Know examples of where they occur, and relate them to your case studies of volcanoes and earthquakes.

Plate margins can be divided into three types (see Table 7.1):

1 **Destructive margins** (or compressional or convergent margins)
   i subduction zones   ii collision zones

2 **Constructive margins** (or extensional or divergent margins)

3 **Conservative margins** (or tensional or transform margins)

Table 7.1 Types of plate margins

Type of margin	Description of movement	Tectonic activity	Examples (see Figs 7.2, 7.12 and 7.13)
1 Destructive i Subduction zones (Fig 7.4)	• Ocean and continental plates converge • Denser oceanic plate sinks beneath the continental plate forming an **ocean trench** and **subduction zone** • Friction between the plates, and heat from the mantle, melts the oceanic plate which becomes molten magma • Under pressure, magma rises through the crust and forms a volcano • Friction and pressure between plates causes earthquakes • Pressure pushes the rocks upwards to form **fold mountains** • Where volcanic activity and fold mountains form off shore, an **island arc** is created	Large, violent earthquakes Violent volcanic activity Some of the largest volcanoes in the world Japan has 60 active volcanoes and frequent earthquakes	Nazca/ South American plates Japan Philippine /Eurasian plates Caribbean islands Caribbean / North American plates
ii Collision zones (Fig 7.5)	• Two continental plates collide • Equal density; neither plate sinks • Rock is compressed and forced upwards to form **high fold mountains**	Many earthquakes; sometimes violent	Himalayas Indo-Australian plate / Eurasian plate **Alps** (see Fig 7.8) Andes Zagros Mountains (south Iran)
2 Constructive (Fig 7.6)	• Plates move apart • Molten rock (magma) rises to the surface to fill the gap • Cools and forms **new** oceanic crust • Forms mid-oceanic ridges and volcanoes on the ocean floor • If a ridge or volcano rises high enough it will emerge at the surface to form a new island	Gentle volcanic activity Weak earthquakes	Mid-Atlantic Ridge Islands of Iceland, Surtsey (1963), Azores
3 Conservative (Fig 7.7)	• Plates move past each other, in similar or opposite directions • No crust is created or destroyed • Pressure builds up over time and is released in a sudden jerk • Violent shock waves are released causing earthquakes	Frequent, sometimes violent, earthquakes	San Andreas Fault (California) Pacific / North American plates

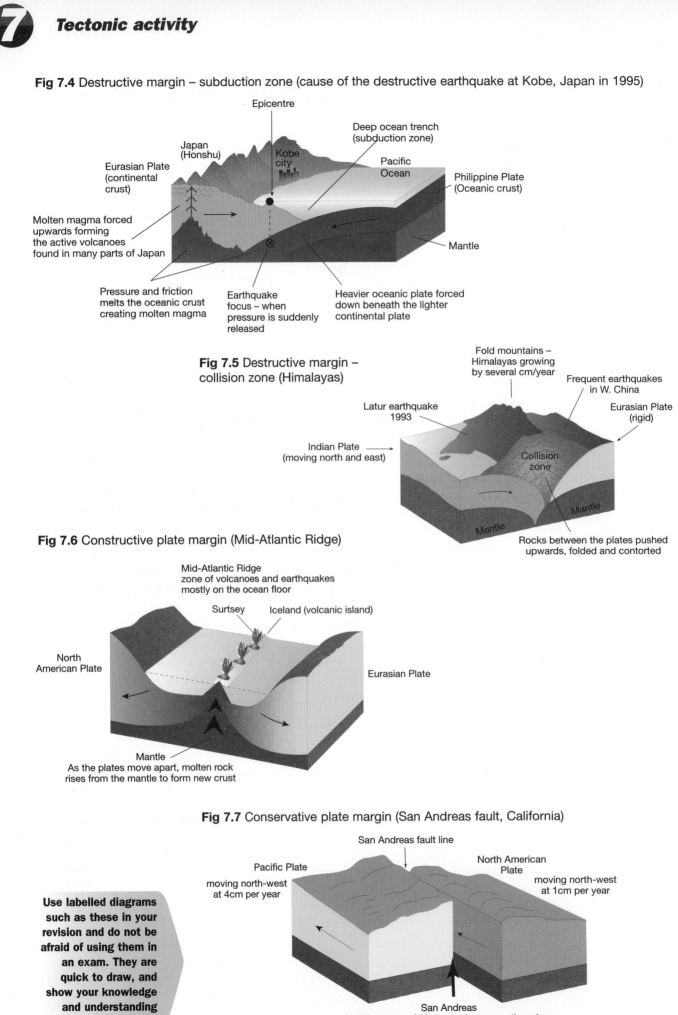

Fig 7.4 Destructive margin – subduction zone (cause of the destructive earthquake at Kobe, Japan in 1995)

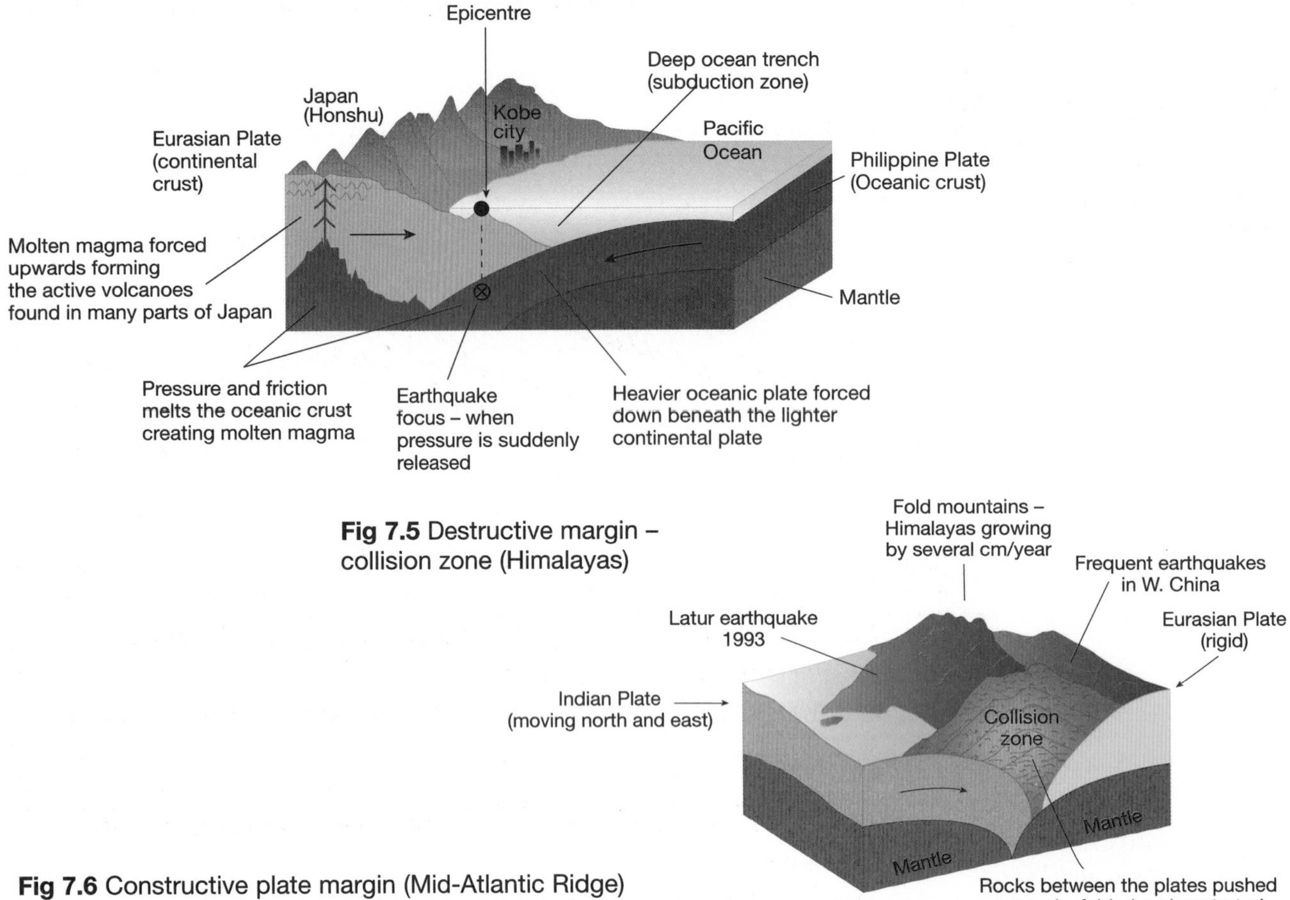

Fig 7.5 Destructive margin – collision zone (Himalayas)

Fig 7.6 Constructive plate margin (Mid-Atlantic Ridge)

Mid-Atlantic Ridge zone of volcanoes and earthquakes mostly on the ocean floor

Surtsey

Iceland (volcanic island)

North American Plate

Eurasian Plate

Mantle

As the plates move apart, molten rock rises from the mantle to form new crust

Fig 7.7 Conservative plate margin (San Andreas fault, California)

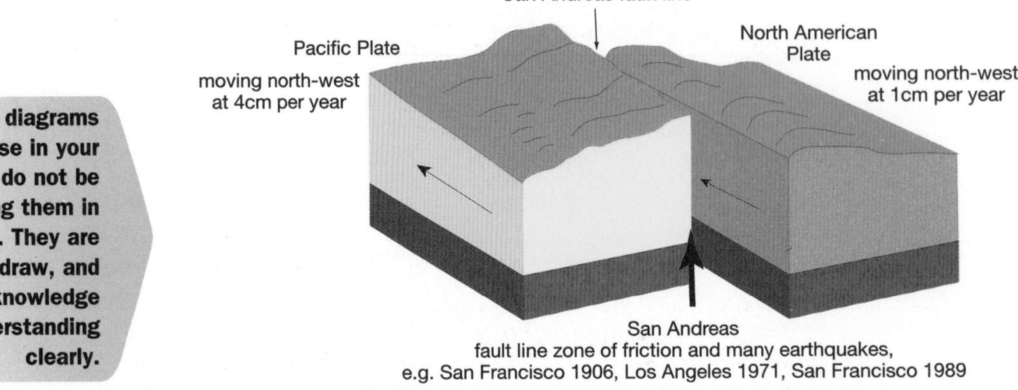

Use labelled diagrams such as these in your revision and do not be afraid of using them in an exam. They are quick to draw, and show your knowledge and understanding clearly.

Formation of fold mountains

AQA A

Case study – The Alps

How are rocks containing marine fossils found high up in the mountains?

Fig 7.8 Diagrammatic sketch showing formation of the Alps

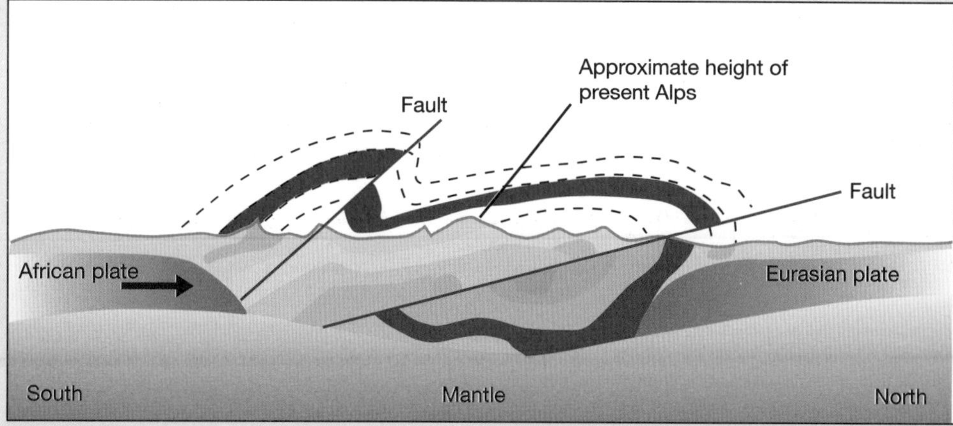

- The African plate moved towards the rigid European plate (Destructive Margin – collision zone – see page 97).
- Between the two plates were thick beds of weak sedimentary rocks which had been laid down in a deepening sea.

This sequence is important. Draw and label the diagram to help you understand it.

- As the plates converged, the sedimentary rocks were squeezed upwards into huge folds.
- More compression led to complex folding and faulting.
- The movements created great pressure and heat which changed some of the sedimentary rocks into much harder metamorphic slates and marbles.

- Weathering and erosion by rivers and ice has produced the landscape of high mountain ranges and deep river valleys we see today.

Fig 7.9

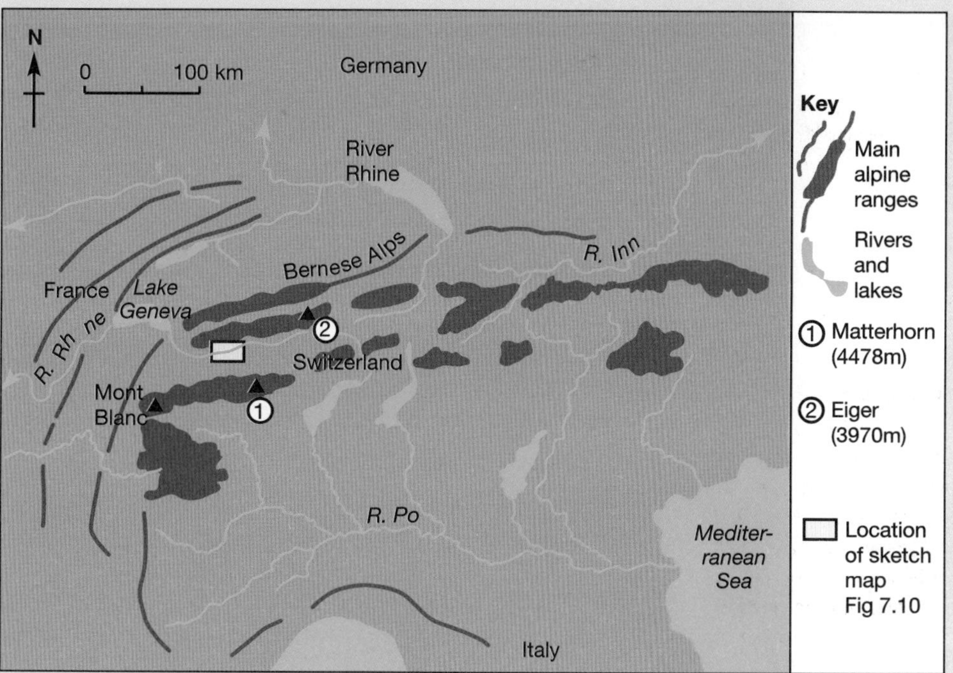

- The tectonic activity has resulted in a complex geology which can give rise to dramatic changes in height over short distances – compare the heights of Mont Blanc (4807m) and Lake Geneva (321m), which are only 70km apart.
- The river valleys are separated by high mountain ranges, e.g. Bernese Alps.
- Major rivers such as the Rhône and Rhine have their sources in the Alps. Both rivers emerge as meltwater from beneath large glaciers.
- Both rivers follow deep steep-sided, flat floored 'glacial troughs' (see page 48).

Human activity in young fold mountains

Case study – The Swiss Alps – the Upper Rhône Valley

Fig 7.10 Sketch cross-section of Upper Rhône Valley near Sion

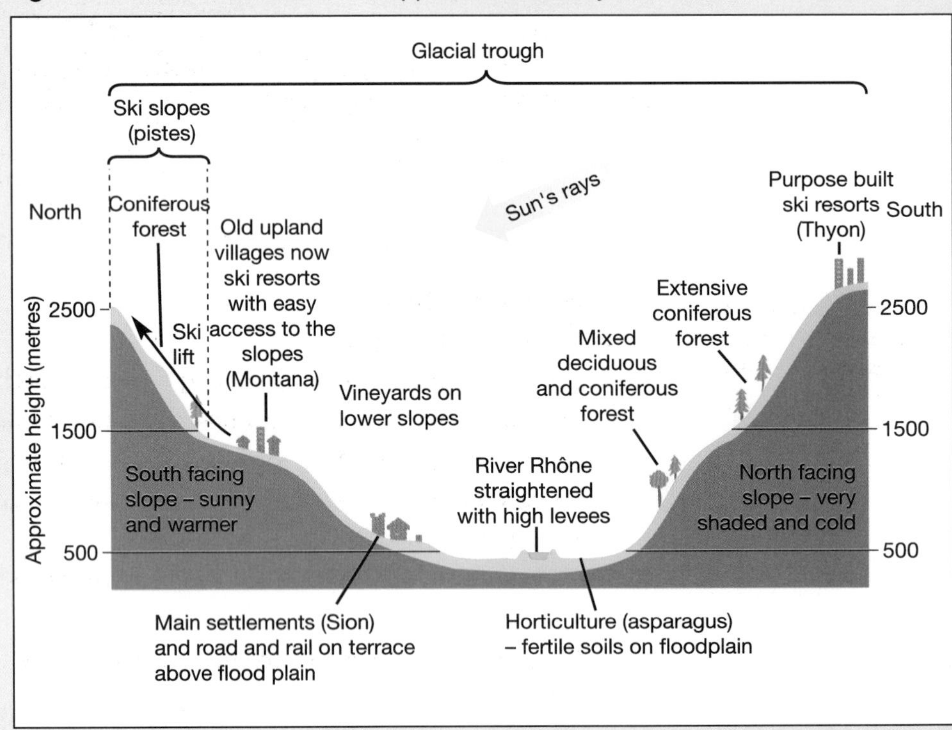

You need to be able to explain why and how these features were formed.

Physical features

- High snow-capped mountain ranges (Fig 7.9: Bernese Alps, Mont Blanc Massif).
- Deep glacial troughs (Rhône valley) with hanging valleys (waterfalls and alluvial fans), and truncated spurs.
- Cirques and arêtes (Wildhorn – Arpelistock Ridge) (Fig 7.11).
- Many rapidly flowing rivers, with steep rocky courses.
- Large rivers flow across flood plains in the wide flat-floored glacial troughs (Rhone).
- Compare the narrow, deep V-shaped hanging valleys to the Rhône valley cross section (Fig 7.11)
- Deep alpine valleys have contrasting local micro-climates on the north and south facing slopes, which can affect human activities (Fig 7.10).

Fig 7.11 Sketch map of Upper Rhône Valley

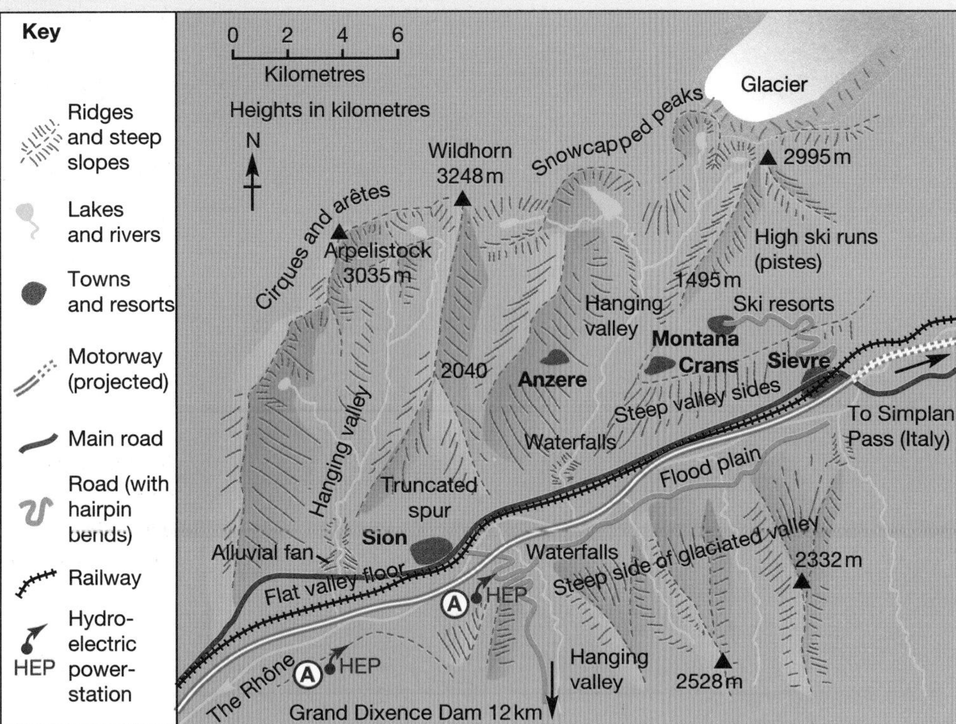

Give examples and explain where these activities are found.

Human activities

- Winter sports, e.g. skiing.
- Summer tourism (touring, walking based in hotels in towns such as Sion).
- Agriculture and forestry – horticulture and vineyards.
- Communication routes – main roads and railways follow main valleys – tunnels and high passes, e.g. Simplon Pass and Gotthard tunnel, provide important links with Italy.
- Rivers such as the Rhone have been straightened and embanked to prevent flash floods.
- Dams and hydro-electric power stations provide electricity.

Investigate examples of these on the internet.

Hazards in alpine areas

- Flash flooding from heavy rain or rapid snow melt.
- Convectional storms with thunder and lightning.
- Avalanches, mudslides and rock falls.

See also Chapter 11, page 145.

KEY POINT

Hydro-electric power (HEP) accounts for 60% of Switzerland's electricity production. One third of this power (20% total) is generated at the power stations marked A and A (Map Fig 7.11). They produce nearly 2000MW (cf. Heysham 2 (UK) 1250MW). The turbines are driven by water falling through pipes from a reservoir in a mountain valley 1500m above (the head). The deep, narrow hanging valleys provide good dam sites. The Grand Dixence Dam is 285m high (one of the highest in Europe) and only 700m wide (top), yet it stores water in a reservoir 5km long (600 million cu.m). Electricity is used in Switzerland to power the railways and urban public transport as well as serving industrial and domestic demands.

Where are earthquakes and volcanoes located?

AQA A AQA C

Questions often ask you to describe the 'pattern' shown on a map. These two maps are good for practising your technique.

There is a clear pattern for the distribution of both earthquakes and volcanoes. The patterns are closely linked to the distribution of plate margins (Fig 7.2).

Volcanoes occur:

- **in narrow bands; a pattern which is similar but not identical to earthquakes distribution, e.g. volcanoes are not found along collision margins (Himalayas)**
- **most often at destructive plate margins, e.g. around the Pacific Ocean – the 'Ring of Fire'**
- **along constructive plate margins, e.g. Mid-Atlantic Ridge**
- **at 'hot spots' in the centre of plates where the crust is thin, e.g. Mauna Loa, Hawaii.**

Fig 7.12 Global distribution of volcanoes

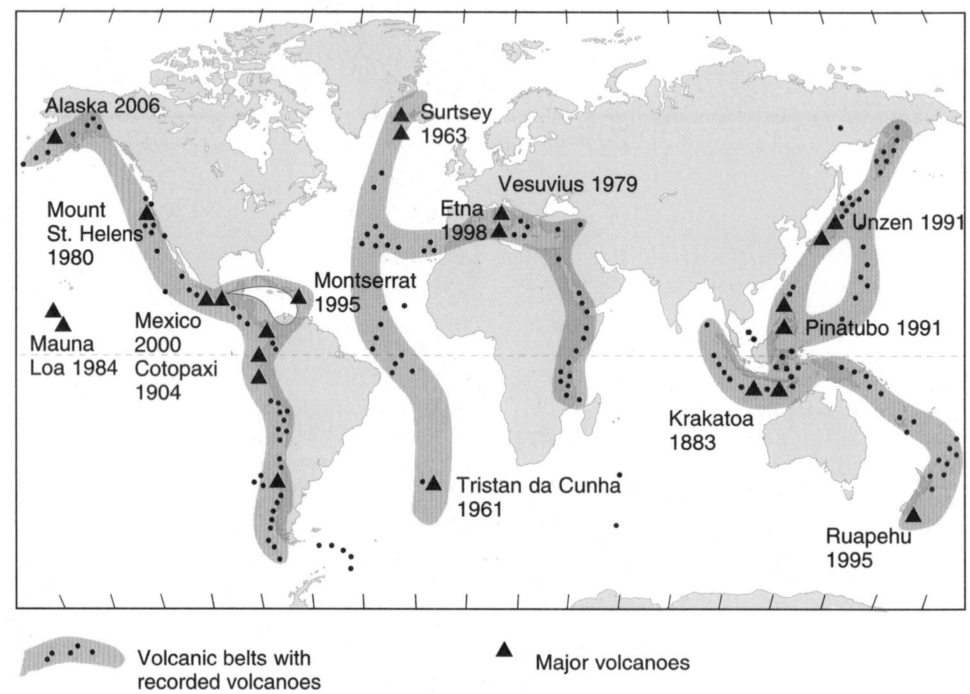

Earthquakes occur:

- **in long narrow bands**
- **along all types of plate margin**
- **close to volcanoes**
- **on land or under water**
- **in a dense ring around the Pacific Ocean.**

Fig 7.13 Global distribution of earthquakes

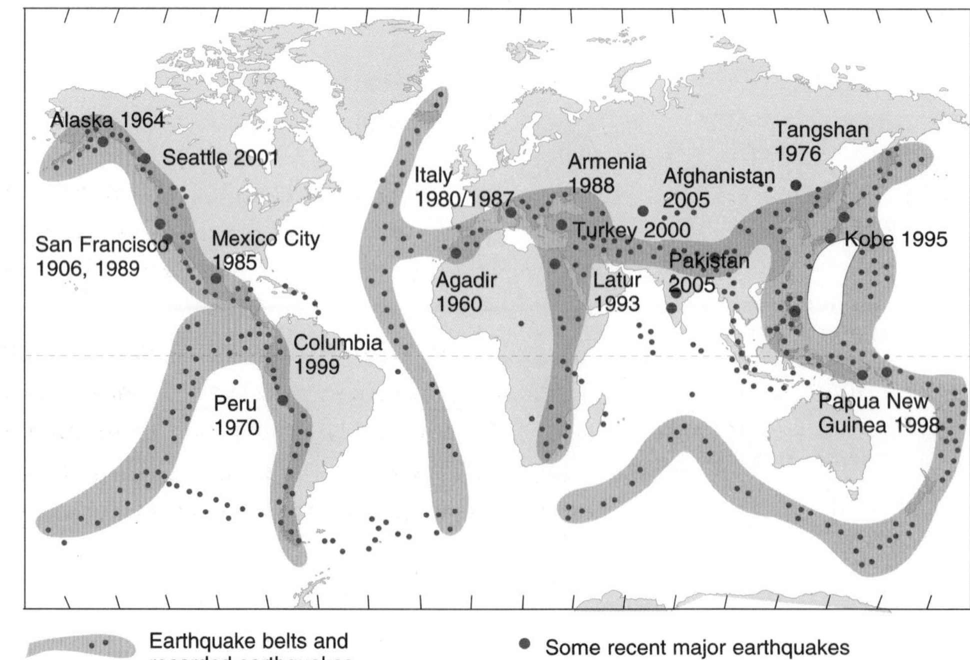

Volcanoes

AQA A AQA C

Volcanoes occur where weaknesses in the Earth's crust allow magma, ash, gas and water to erupt onto the land and seabed.

Fig 7.14 Characteristics of a composite volcano (Soufriere Hills volcano, Montserrat in the Caribbean)

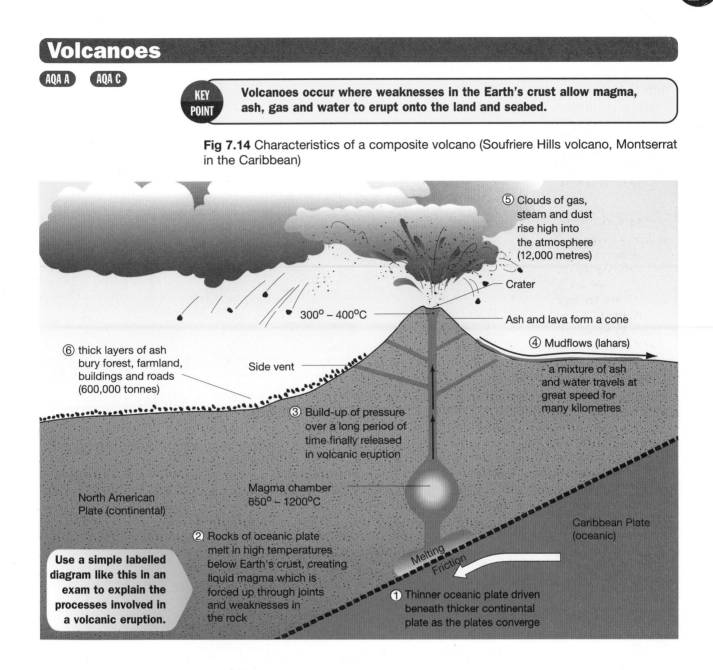

⑤ Clouds of gas, steam and dust rise high into the atmosphere (12,000 metres)

Crater

300° – 400°C

Ash and lava form a cone

④ Mudflows (lahars) – a mixture of ash and water travels at great speed for many kilometres

⑥ thick layers of ash bury forest, farmland, buildings and roads (600,000 tonnes)

Side vent

③ Build-up of pressure over a long period of time finally released in volcanic eruption

Magma chamber 650° – 1200°C

North American Plate (continental)

② Rocks of oceanic plate melt in high temperatures below Earth's crust, creating liquid magma which is forced up through joints and weaknesses in the rock

Melting

Friction

Caribbean Plate (oceanic)

① Thinner oceanic plate driven beneath thicker continental plate as the plates converge

Use a simple labelled diagram like this in an exam to explain the processes involved in a volcanic eruption.

Types of volcano

Most volcanoes occur at plate margins.

Volcanoes can also be classified in the following way (locations on Fig 7.12):

- **Active**: volcanoes that have erupted recently, e.g. Soufriere Hills, Montserrat; Mount St Helens, north west USA; Augustine volcano, Alaska (2006).
- **Dormant**: volcanoes that have not erupted for a long time, e.g. Mount Rainier, north west USA. Soufriere Hills volcano in Montserrat had been dormant for 400 years before the violent eruption in 1996.
- **Extinct**: volcanoes that are no longer areas of tectonic activity; no record of volcanic activity since records began, e.g. Castle Rock, Edinburgh.

Dormant volcanoes always pose an element of risk for the people living and farming nearby.

Fig 7.15 Types of volcano

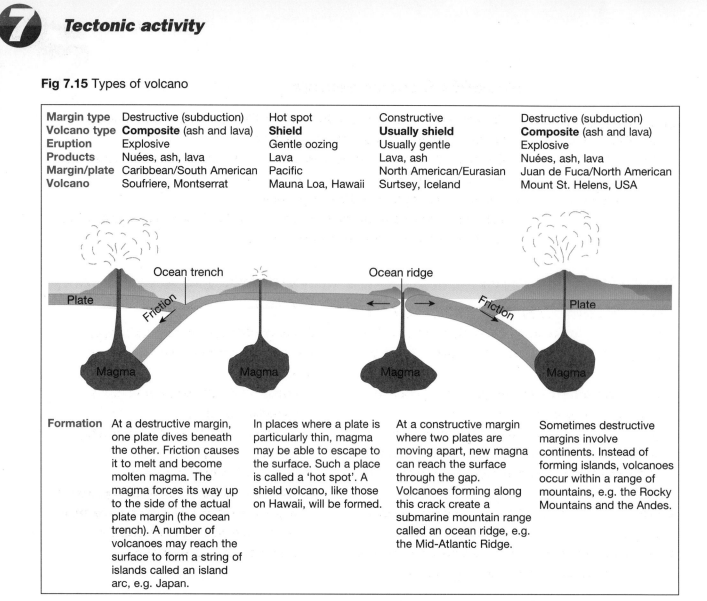

Margin type	Destructive (subduction)	Hot spot	Constructive	Destructive (subduction)
Volcano type	**Composite** (ash and lava)	**Shield**	**Usually shield**	**Composite** (ash and lava)
Eruption	Explosive	Gentle oozing	Usually gentle	Explosive
Products	Nuées, ash, lava	Lava	Lava, ash	Nuées, ash, lava
Margin/plate	Caribbean/South American	Pacific	North American/Eurasian	Juan de Fuca/North American
Volcano	Soufriere, Montserrat	Mauna Loa, Hawaii	Surtsey, Iceland	Mount St. Helens, USA

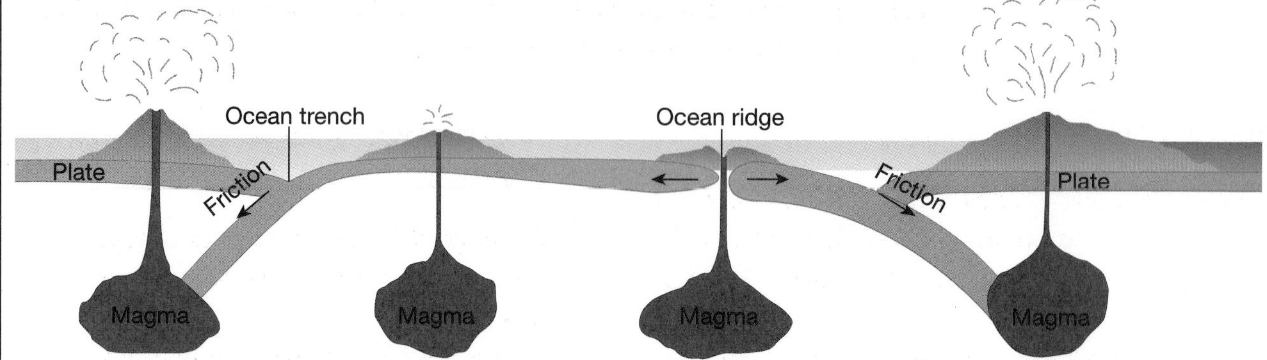

Formation	At a destructive margin, one plate dives beneath the other. Friction causes it to melt and become molten magma. The magma forces its way up to the side of the actual plate margin (the ocean trench). A number of volcanoes may reach the surface to form a string of islands called an island arc, e.g. Japan.	In places where a plate is particularly thin, magma may be able to escape to the surface. Such a place is called a 'hot spot'. A shield volcano, like those on Hawaii, will be formed.	At a constructive margin where two plates are moving apart, new magna can reach the surface through the gap. Volcanoes forming along this crack create a submarine mountain range called an ocean ridge, e.g. the Mid-Atlantic Ridge.	Sometimes destructive margins involve continents. Instead of forming islands, volcanoes occur within a range of mountains, e.g. the Rocky Mountains and the Andes.

Hazards resulting from volcanoes

Table 7.2

Hazard	Description	Examples
Lava flows	Molten rock (magma) flowing down the sides of a volcano. Very hot basaltic lava flows very fast.	Frequent rapid lava flows on Mt Etna, Sicily, threaten local settlements.
Lahars	Mudflows, a mixture of ash and water, travel at great speed down the mountain sides.	Ruiz volcano, Columbia, erupted in 1985. The eruption melted snow around the summit and a mixture of water, ash and rock turned to mud and buried the town of Armero. 22 000 people drowned in the mud.
Dust and ash clouds	Ash thrown high into the atmosphere shuts out the sun and causes gloomy days. On settling, ash can completely bury buildings and crops.	When Soufriere Hills volcano in Montserrat erupted in 1996, ash clouds deposited 600 000 tonnes of ash burying the landscape and farms.
Nuée ardente (glowing cloud)	Flows of superheated ash and gases, flowing at very fast speeds.	In 1980, nuée ardente from the eruption of Mt St. Helens in north-west USA destroyed every tree over a very large area (see Fig 7.13).

Benefits from volcanoes

These benefits help to explain why large numbers of people live in the danger areas surrounding volcanoes.

- Volcanic ash weathers to produce very fertile soils, excellent for intensive farming, e.g. lower slopes of Mt Etna, Sicily.
- Geothermal energy – Iceland uses energy from volcanic water and steam to supply heat and electricity.
- Volcanic rock makes good building stone.
- Volcanoes attract adventurous tourists to visit sites in Iceland, New Zealand and southern Italy (Vesuvius and Stromboli).

Predicting volcanoes

Careful monitoring has allowed scientists to become more proficient at predicting volcanic eruptions. Early evacuation can be advised and lives saved. The main methods include:

- earthquake activity increases prior to eruption as the magma rise in the crust
- gas emissions change with increasing sulphur
- lasers and tiltmeters measure changes in the shape of the volcano; bulges and domes suggest a build up of magma
- ground temperature changes can be measured by satellite
- ultrasound is used to detect movements of magma
- studies of past activity may provide evidence.

Responding to a volcanic hazard

Although little can be done to control volcanoes, some measures that can be taken are:

- lava flows have been diverted from vulnerable areas
- lava flows have been halted by water sprays
- design buildings to shed ash
- communities can prepare for a disaster:
  - prevent settlement in vulnerable areas
  - organise evacuation procedures
  - improve public awareness.

LEDC case study: Montserrat, Caribbean, (British Territory) 1995-97 (see Fig 7.12)

Fig 7.16 Montserrat: eruptions 1995-1997

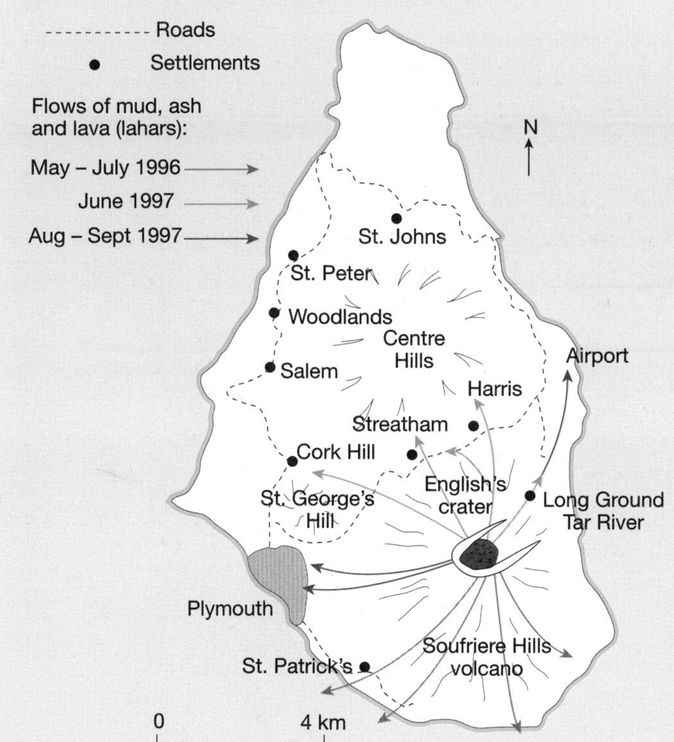

Montserrat factfile:

Population: 11 000 (1995)
An island in the island arc on the destructive plate margin between the North American and Caribbean plates.

Effects of the eruption

- Forests and rich farmland destroyed by lahars and covered with deep layers of ash.
- Many villages buried in ash; 23 people killed.
- Many people evacuated:
  - to the north of the island
  - to neighbouring islands
  - to the UK.
- The capital city, Plymouth, (port, industrial and administrative centre) almost totally destroyed and abandoned.
- The only hospital was destroyed.

- The only airport was burned and engulfed in ash.
- Many roads were destroyed.
- Dust covered the whole island, making it difficult to breathe.

The effects of volcanoes and earthquakes are more severe and long-lasting in LEDCS than in MEDCs. Be prepared to explain why.

Short term responses

- The authorities and people were totally unprepared since the volcano had been dormant for 400 years.
- People evacuated to the north of the island, were housed in tents and makeshift homes with little food, poor sanitation and no power.
- The UK government gave £55 million in compensation and redevelopment and 250 prefabricated houses.
- The hospital reopened in a former school.
- Communications were difficult and expensive to repair.
- Many people are unemployed since the tourist industry collapsed.
- Little farmland to reclaim in the south.
- Plymouth – capital city abandoned.

Long term responses

- The volcano is still active with a new lava dome growing in 2006. A volcanic observatory has been established to monitor activity and advise residents and tourists of the dangers.
- An exclusion zone exists over much of the southern part of the island and for two kilometres off-shore.
- The population is about 5000, under half that in 1995. All reside in the south of the island, where few lived before the eruption.
- There is no capital city. The ash covered remains of Plymouth remain in the exclusion zone.
- Services have been expanded in the north – there are reliable electricity and water supplies. Education exists for all over five years.
- New roads have been built in the south and are being extended.
- Agriculture and fishing once employed nearly 500 people, now it is only 200. The government is concentrating on poultry production and other methods to reduce the island's dependency on imports.
- Construction of an industrial park is in progress.
- Tourism is seen as the major enterprise in the future, outside investment is encouraged but growth is slow. The tourist department advertises 'volcano experiences', 'diving and snorkelling', 'weddings and honeymoons' as attractions – international access is only via Antigua and a one-hour ferry trip.

MEDC case study: Mt St. Helens, north-west USA, 1980 (see Fig 7.12)

Fig 7.17 Effects of Mt St Helens eruption

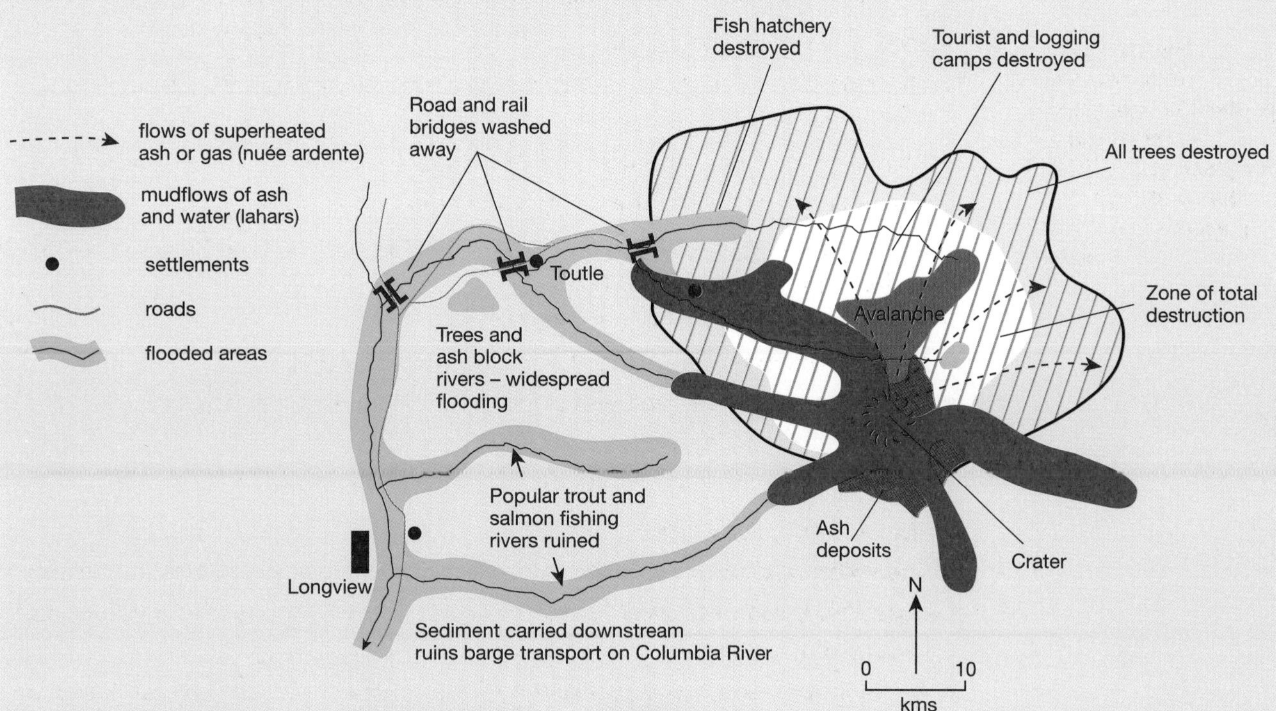

Compare the response of the USA (an MEDC) with that of Montserrat (an LEDC) to a volcanic event.

Mt St. Helens factfile

Mt St. Helens lies on the destructive margin where the Juan de Fuca plate (oceanic) is disappearing below the North American plate.

Effects of the eruption

- Extensive forests and logging camps destroyed by nuée ardente.
- Ash and flooding destroyed crops and livestock.
- Lava flows and ash clogged prime salmon and trout rivers.
- Early warnings were given as scientists monitored the volcano; most people evacuated to safety.
- 63 people were killed, most from poisonous gases.
- Flooding washed away road and railway bridges.

Response to the eruption

- Ten million trees were replanted.
- Agriculture benefited from fertile ash.
- Rivers were dredged and restocked with fish.
- People were re-housed.
- Bridges were rapidly rebuilt.
- Tourism has increased, attracted by the event. A new tourist centre has been built, attracting a million visitors a year. New tourist lodges have also been put up.
- The volcano is still active and carefully monitored by scientists.

Volcanic activity still continues and is carefully monitored at both case study locations. Research the latest information.

Earthquakes

AQA A AQA C

An earthquake is the vibration of the Earth's crust caused by shock waves travelling outwards from a sudden movement deep within the crust.

The source of the shock waves is known as the **focus**, and the point on the Earth's surface immediately above is the **epicentre**. Most earthquakes are associated with movements along plate margins, but many occur at other weaknesses in the crust such as fault-lines.

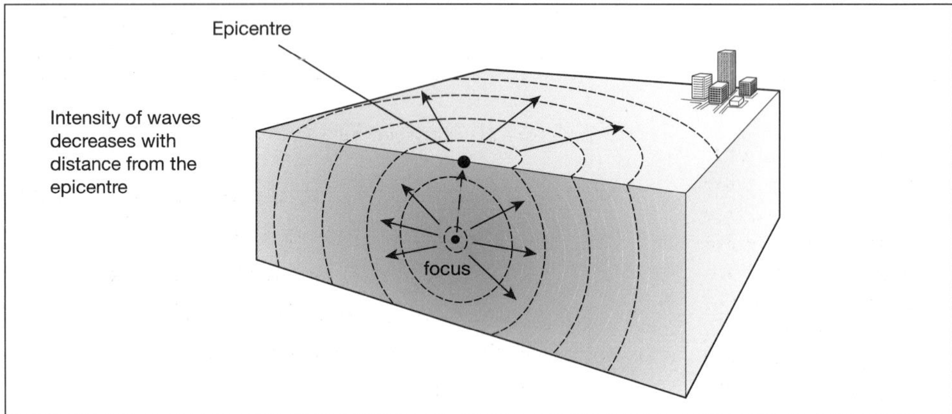

Fig 7.18 Characteristics of an earthquake

Measuring earthquakes

The shock waves of an earthquake are recorded and measured on a **seismograph**. The seismograph measures the strength or magnitude on the **Richter Scale** (see Fig 7.19).

Add examples of more recent earthquakes to Fig 7.19.

Fig 7.19 The Richter Scale

Earthquake size (magnitude measured by seismograph)	Magnitude	Possible effects	Number
	1	None, only detected by instruments	100,000 per year
	2	Faint tremor, like the vibrations of a passing lorry	c 100 per year
	3		
	4	Structural damage to chimney pots. Trees sway	
North Wales 1984 (largest UK earthquake) 5.4	5	Distinct shaking, poorly built houses collapse. Ground cracks	c 15 per year
Bam, Iran 2003 6.7	6		
Seattle 2001 7.0; Kobe 1995 7.2	7	Major earthquake, large buildings and infrastructure destroyed	
Gujarat, India 2001 7.9; Mexico City 1985 8.1	8	Widespread destruction	1 every 5 years
Chile 1960 8.9; Banda Aceh (Sumatra) 2004 9.0 +	9		

The impact of an earthquake

'The greater the strength of an earthquake, the higher the death toll'. Is this statement correct? Use Table 7.3.

- Primary impact includes:
  - deaths
  - collapse of buildings
  - damage to roads, water and gas.
- Secondary impact includes:
  - fire and flooding
  - people made homeless
  - injuries and disease
  - loss of business

Table 7.3 The impact of individual earthquakes

Year	Place (see Fig 7.9)	Richter Scale	LEDC/MEDC	Impact
1960	Agadir, Morocco	5.7	LECD	12 000 deaths, many homeless
1976	Tangshan, China	7.8	LEDC	250 000 deaths, 650 000 homeless
1989	San Francisco, USA	7.1	MEDC	62 deaths
1994	Los Angeles, USA	6.7	MEDC	57 deaths, 20 000 homeless
1999	Turkey	7.3	LEDC	18 000 deaths, many remote villages destroyed
2001	Gujarat, India	7.9	LEDC	100 000 deaths, 30 000 homeless
2001	Seattle, USA	7.0	MEDC	1 death, 250 injuries, damage: billions of dollars
2005	Pakistan	7.6+	LEDC	100–150 000 dead countless numbers homeless in high mountains

Tsunamis are not tidal waves. In December 2004 a submarine earthquake occurred along the destructive plate margin west of Sumatra (Indian Ocean). An upward thrust of 20m of the seafloor sent away a series of giant waves in all directions, moving very fast. In 15 minutes waves 20m high submerged the coast of W.Sumatra ripping vegetation from mountainsides some 800m inland. The city of Banda Aceh was destroyed killing tens of thousands of people. In two hours similar waves hit Sri Lanka, still with no warning; 4000 people died in the region around the city of Galle alone.

Factors affecting the impact of an earthquake

- The strength of the earthquake measured on the Richter Scale.
- The distance from the epicentre.
- The nature of the surface rocks – soft rocks can absorb water and become very mobile, e.g. Mexico City 1985.
- Population density: densely populated urban areas are particularly vulnerable, e.g. Kobe 1995.
- The remoteness of an area will affect the time that emergency services and aid take to reach the area, e.g. Afghanistan 1998, Pakistan 2005.
- Resources available and preparedness of an area; MEDCs are able to afford better services.
- Time of day.
- Time of year/climate can affect survival and the spread of disease.
- Coastal location with regard to impact of tsunamis.

Tsunamis (Japanese for 'harbour wave') are powerful ocean waves caused by earthquakes. They can travel long distances across oceans very quickly. As they approach the shallower water along a coast, the wave suddenly builds up to a great height. The speed, height and power of the wave can cause severe damage and many deaths in low-lying coastal areas.

Responding to earthquakes

Control

Research in the USA has experimented with injecting fluids along the San Andreas fault to lubricate zones of friction (see Fig 7.7). There is little evidence that this works.

Despite much research, it is still not possible to predict earthquakes.

Prediction

It is possible to predict where earthquakes are likely to happen but not when. On-going research is investigating:

- evidence from past earthquakes
- patterns of small tremors
- unusual animal behaviour
- gas emissions from the ground.

One of the reasons for the low death toll and injuries in Seattle (2001) was that large sums of money had been spent over the last 20 years to ensure buildings can withstand major shocks. The Space Needle, built to withstand a 9.1 magnitude earthquake, shuddered violently but no damage occured.

Preparation

Preparation for earthquakes and their aftermath can include:

- land use planning to prevent building on weak rock, e.g. reclaimed land from the sea
- improved building design, including:
  - rolling weights on roof to counteract the shock waves
  - identification number on roof visible to helicopters assessing the damage after an earthquake
  - 'bird cage' interlocking steel frames
  - automatic shutters that come down over the windows to prevent pedestrians being showered with glass
  - panels of marble and glass flexibly anchored into the steel superstructure
  - rubber shock absorbers between the foundations and superstructure
  - reinforced foundations deep in bedrock
  - open areas outside buildings where pedestrians can assemble if evacuated
- more flexible gas, water and power lines
- planned emergency and evacuation procedures (regular earthquake drills in Japan)
- greater public awareness – education in schools, e.g. Japan and California.

MEDC case studies: Kobe, Japan, 1995

Kobe factfile

Location: Kobe, Japan (see Fig 7.13); a heavily populated urban area

Time and date: 5.46 a.m. on 17th January 1995

Strength on Richter Scale: 7.2

Plate margin: destructive margin – subduction zone (see Fig 7.4)

Short-term effects

- 5477 deaths
- 35 000 injured
- 316 000 left homeless
- Many houses collapsed despite earthquake-proof designs.
- Large number of buildings destroyed by fire when gas mains fractured.
- Many people evacuated to schools, parks and community centres; tents and emergency rations provided.
- Rescue teams arrived quickly on scene with equipment.
- National government set up emergency headquarters in two days.
- Injured people transferred to hospitals in nearby cities.
- Electricity restored in six days.
- Search for survivors concluded in ten days.
- Earthquake refugees started moving into temporary housing in two weeks.
- Much industry, communications and shopping restored within three months.
- All temporary shelters for homeless closed in seven months.

Long-term effects

- Many businesses closed.
- Extra jobs created in rebuilding.
- Many people moved away from area permanently.
- Overall cost 10 billion yen.

LEDC case study: Latur, Maharashtra, central India, 1993

Latur factfile

Location: Latur, Maharashtra, India (see Fig 7.2); a well-populated, remote, rural area
Time and date: 4.00 a.m. in September 1993
Strength on Richter Scale: 6.4
Plate margin: Destructive margin – collision zone (see Fig 7.5)

Be prepared to explain why the effects are far more severe in LEDCs than in MEDCs.

Short-term effects

- 22 000 deaths
- 4800 injured
- 10 000 left homeless
- Nearly all the stone houses collapsed, bringing rock and concrete down onto sleeping people.
- Large numbers of people camped in makeshift shelters made from tin sheeting, straw and polythene.
- Army rescue teams did not arrive for 36 hours.
- Large numbers of dead posed threat of disease and polluted water supplies.
- Lack of hospital service; too remote to transfer to other towns.
- Rescue and retrieval of bodies continued for many weeks.
- Earthquake refugees still in camps with poor sanitation many months after the event.

Long-term effects

- Very little temporary housing was provided.
- Some villages remain in ruins.
- Farming, which relies on hand labour, was devastated by loss of life.
- Families were left destitute.
- International aid agencies were mainly responsible for rehabilitation.
- Indian government asked for aid to meet the £22 million costs of rescue.
- Long-term aid for the region was insufficient.

1. What are the two types of plate forming the Earth's crust?
2. What type of plate boundary is causing the formation of the fold mountains of the Himalayas?
3. What type of plate boundary is found close to Japan, at the boundary between the Eurasian and Pacific Plates?
4. What type of plate boundary is found along the Mid-Atlantic Ridge?
5. What is the name given to the molten rock, which flows down the sides of some volcanoes?
6. How is tectonic activity of benefit to Iceland?
7. What is the name given to the point on the Earth's surface immediately above the focus of an earthquake?
8. What scale is used to measure the strength or magnitude of an earthquake?

1. Oceanic/continental 2. Destructive/collision zone 3. Destructive/subduction zone
4. Constructive 5. Lava 6. Tourism, geothermal energy 7. Epicentre 8. Richter Scale

Chapter

8 Population

LEARNING SUMMARY

After studying this section you should be able to:

- ***describe and understand the main features of the world pattern of population density***
- ***understand the main reasons for population change and the effect on population structure***
- ***describe a range of population issues and their planned solutions***

The world pattern of population density

AQA A AQA B AQA C

KEY POINT

Population density is measured as the number of people living in an area. It is usually measured in people per square kilometre.

People are not evenly spread across the Earth's surface; see Table 8.1.

Table 8.1

Be sure to know some examples of places with different population densities.

Scale	Dense	Sparse
International	SE Asia, Western Europe	Sahara Desert, Antarctica
National	Singapore, Netherlands	Australia, Mongolia
Regional	SE England, SE Brazil	Western Australia, Amazonia
Local	Hong Kong, Rio de Janeiro	Snowdonia, Alaska

Fig 8.1 World population distribution

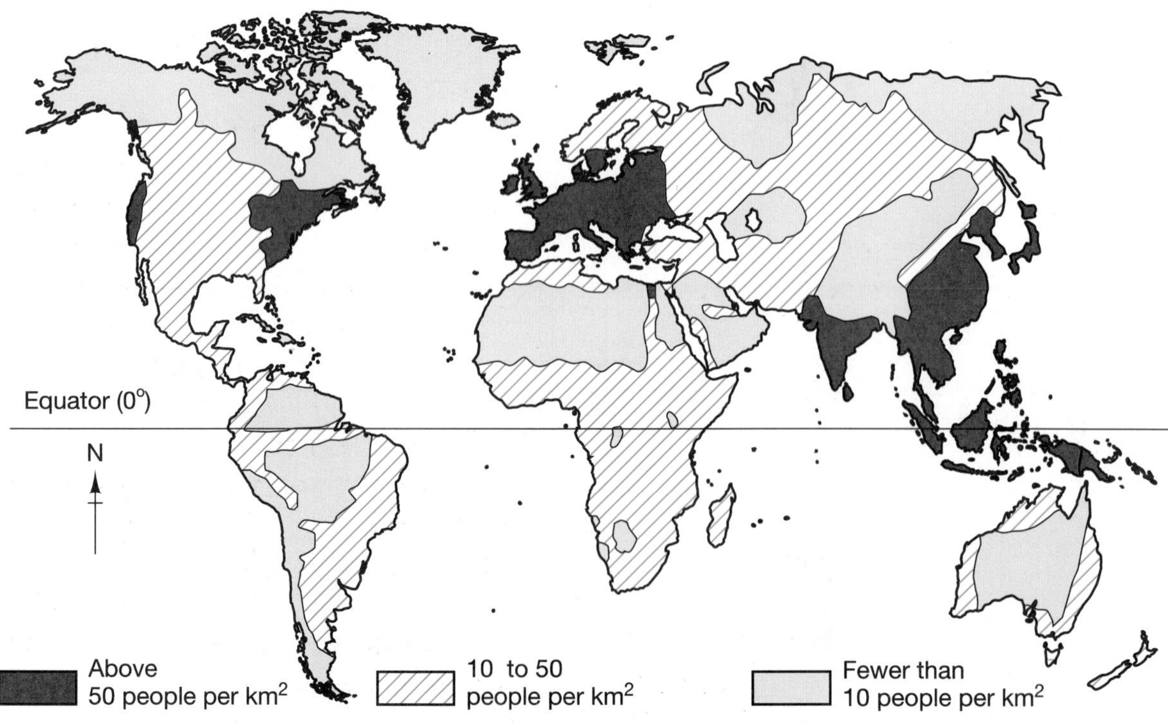

Reasons for the differences in population distribution and density

The factors affecting the distribution and density of population include:

- physical factors such as relief, climate, vegetation, resources
- social factors such as the proportion of urban to rural population
- economic factors such as the resources, amount of industrialisation
- political factors such as government policy on birth control.

Population distribution describes the pattern of people in an area (how the population is spaced or spread out). The three main patterns of distribution are uniform or even, nucleated or clustered, and a random pattern.

MEDC case study: population distribution in Japan

This is very helpful for AQA B.

Fig 8.2 Population density in Japan

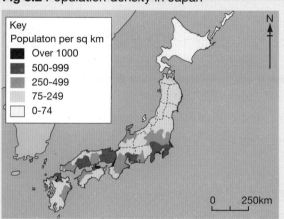

Japan factfile

Area: 380 000 km² (1.5 times the size of the UK)

Population density: 350 people per km² (1.5 times the density of the UK)

Description: Japan is made up of four islands. To the north is the sparsely populated Hokkaido. The largest island, Honshu, is densely populated and contains many large cities, including Tokyo. The northern part of Honshu, and the two southern islands of Shikoku and Kyushu, have intermediate population densities.

LEDC case study: population density in Brazil

Fig 8.3 Population density in Brazil

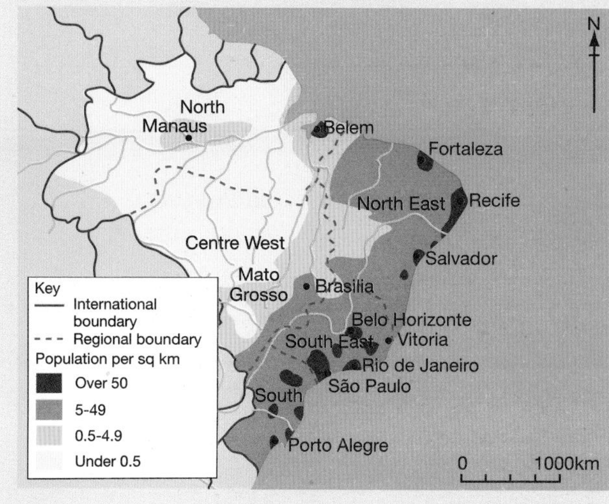

Brazil factfile

Area: 8 511 965 km²

Population density: Brazil has a population density of 20 people per km².

Description: Brazil can be divided into five regions. The North is sparsely populated and contains the Amazon river and a large area of tropical rainforest. The main city is Manaus. Population is increasing as economic development takes place.

The North East contains one in three Brazilians, but its population is declining because of drought.

The South East has the highest population density and includes the cities of Rio de Janeiro and São Paulo. This is the economic centre of Brazil. The South has a high population density with good agriculture and growing industry.

The Centre West area has a low population density but this is growing because the new capital city of Brasilia has been built there.

1. Name two areas of high and two areas of low population density.
2. What is the difference between population density and population distribution?
3. What factors tend to account for areas of low population density?

1. Examples of high population density: London, Singapore, the Netherlands; low population density: Sahara Desert, Amazonia. 2. **Density** refers to the number of people per km^2; **distribution** is the pattern of population such as nucleated or dispersed. 3. Rainfall, temperature, relief, access.

World population growth

  AQA C

Fig 8.4 World population growth

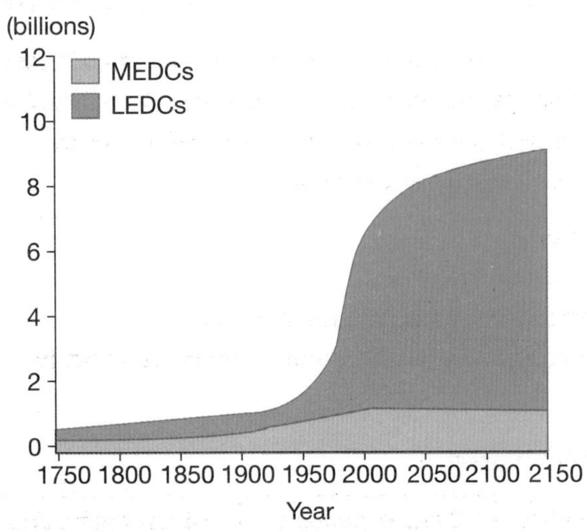

The best estimate of the world's population total is six billion (one billion is equal to one thousand million). The total has doubled since 1960 with most of the growth in LEDCs. Projections of further growth vary. Some experts suggest that the total will reach eight billion. Others believe that social and economic changes in LEDCs will slow their population growth. Many countries take a census (questionnaire survey) every ten years to count their population. A census provides useful information for social and economic planning.

Population change: the balance of birth and death rates

  AQA C

Birth rate is the number of live births per 1000 people per year.
Death rate is the number of deaths per 1000 people per year.

Know one case study of population change in an LEDC and one in an MEDC.

Natural population increase is the difference between **birth** and **death** rates. Natural increase will occur if births exceed deaths.

Birth rates are high in LEDCs such as Bangladesh and India because:

- children provide labour on family subsistence farms
- there are no old age pensions so children provide security for old age
- large families are seen as a sign of the husband's virility
- girls are expected to marry early thus extending their child-bearing years
- women are expected to stay at home and raise a family; they have little education and do not know about birth control
- some religions do not approve of contraception
- high infant mortality encourages large families to ensure some children survive.

Birth rates are low in MEDCs such as the UK, Sweden and Germany because:

- people marry later, women are educated and delay having a family so that they can follow a career
- high costs of living make children expensive
- couples prefer material possessions such as a car, house and holidays

- birth control, and the contraceptive pill in particular, are available
- governments discourage large families to save the costs of building more schools.

Death rates are low in MEDCs and are falling in LEDCs because:

- better healthcare and medical care are available
- people are able to retire earlier after less physically demanding jobs
- cures are being found for diseases such as cancer, malaria and cholera
- people are better educated about hygiene
- there are cleaner water supplies and sanitary facilities
- higher incomes buy the right food, housing and heating.

The Demographic Transition Model

AQA A

Geographers studying the change taking place in the populations of countries have noticed that they follow a similar pattern. This is called the Demographic Transition Model. The model shows how the total population of a country changes through time as birth and death rates change.

The demographic transition model is useful for:

- studying the way population is changing
- understanding trends in births, deaths and natural increase
- predicting the changing structure of population and planning to meet its changing needs.

Fig 8.5 The Demographic Transition Model

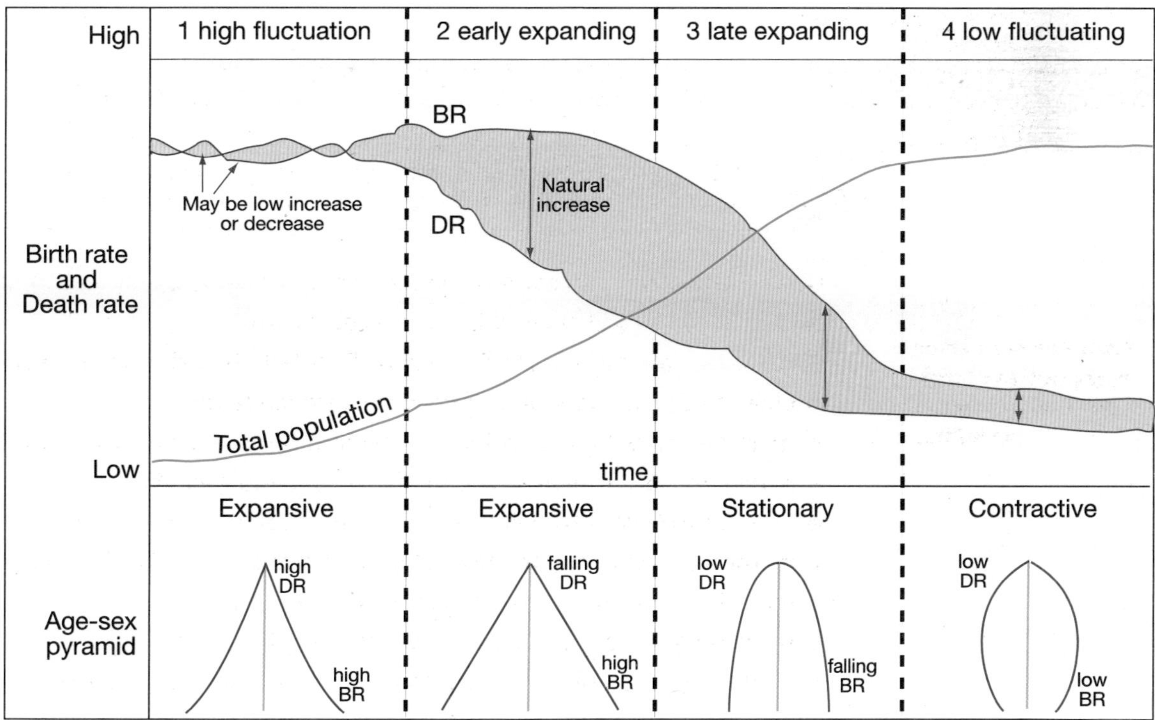

Table 8.2 The Demographic Transition Model

Stage	One	Two	Three	Four
Birth rate	High	High	Falling	Low and varies
Death rate	High and varies	Falls	Low	Low
Population changes	Small	Rapid increase	Slower growth	Stable
Places	Rainforests	Malawi	China	UK, Germany
People's lives	Subsistence agriculture; high infant mortality	Better food supply; control of disease; lack of birth control	Better living conditions and healthcare; growth of industry, jobs	High standards; affluence; small families; long life expectancy; education for all

This is a very popular diagram with examiners. Note how total population is changing at each stage.

Some MEDCs have moved beyond Stage 4. They are described as being in Stage 5. They have low birth rates and low death rates. Their total population may be decreasing. The age-sex pyramid for these Stage 5 MEDCs shows an increasing number of older people and a decreasing number of children.

The limitations of the Demographic Transition Model are:

- LEDCs may not follow the patterns of change found in MEDCs 30 to 50 years ago
- birth rates have not fallen as rapidly as might be expected in some LEDCs because of social customs and beliefs
- government planning for population change may interrupt the model, e.g. the 'One-Child Policy' in China
- some industrialising LEDCs are moving more rapidly through the stages than the MEDCs did.

Divide the pyramid into three age groups.

Population pyramids

The structure of a population can be studied using a population pyramid.

Fig 8.6 Population pyramid for France (MEDC)

Different age groups

Elderly 'retired' (65+)

Working population (15-64)

Children (0-14)

Male

Female

85-89, 80-84, 75-79, 70-74, 65-69, 60-64, 55-59, 50-54, 45-49, 40-44, 35-39, 30-34, 25-29, 20-24, 15-19, 10-14, 5-9, 0-4

Wider at top = larger proportion of elderly; ageing population as life expectancy increases more females than males

Pyramid not tapering = working population has higher life expectancy

Narrow base = smaller proportion of young people; low birth rate

5 4 3 2 1 0 — % of total male population

0 1 2 3 4 5 — % of total female population

Fig 8.7 Population pyramid for Brazil (LEDC)

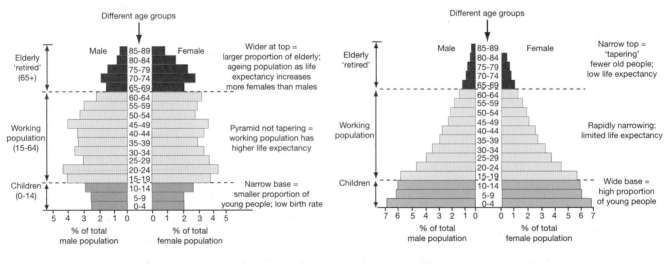

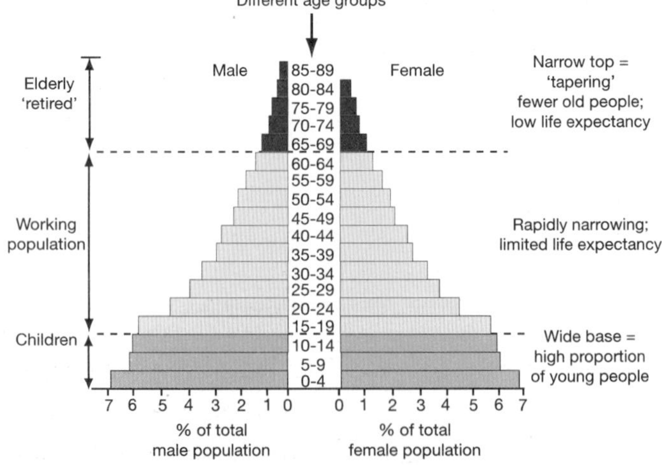

1. Make a list of the reasons for high birth rates in LEDCs.
2. Why are death rates tending to fall in most countries?
3. Name the four stages of the Demographic Transition Model.
4. Why are some MEDCs described as being in Stage 5 of the model?

1. See page 114 2. See page 115 3. See page 115 4. See page 116

Population change

AQA A AQA B AQA C

Population migration

Population totals change with variations in birth and death rates. In addition, population totals change because of migration.

KEY POINT

A migrant is a person who moves to live in another place either permanently or temporarily.
Immigrants move into an area and emigrants move out.

Types of migration

Migration is not part of natural population increase.

There are four types of migration:

- rural to rural, e.g. a farmer retiring to a cottage near the sea
- rural to urban, e.g. people moving from the countryside to cities for jobs and other amenities
- urban to rural, e.g. families moving to a house with a garden in a village from where working members commute back to the urban area
- urban to urban, e.g. people moving from one urban area to another for a job change or moving from a big city to a smaller urban area for better housing.

Another division of migration is into forced and voluntary migrants.

- Forced migrants have had to move, e.g. Jews from the Nazi regime in Germany in 1939.
- Voluntary migrants move because they believe there are benefits if they move.

A model for understanding migration

Fig 8.8 The push–pull model of migration

Push factors	Journey factors	Pull factors
Lack of jobs	Cost of the journey	Job opportunities
Poor housing	Other attractions on the way	Housing
Poor environment, e.g. pollution, crime, traffic, dust, eyesores	Running out of money for travel	Green areas, safety, privacy
	Poor transport	Good schools
War, genocide, persecution	Natural disasters, e.g. floods	Affordable healthcare for all
Poor schools	Long distances	Using wealth to improve life
Lack of healthcare	Language difficulties	Slower pace of life
Retirement	Homesickness	Friends
Family breakdown		People of similar roots
Racial tension		New experiences
Restlessness		New community emerging
Decline of a community		Increased wealth

Case study of rural-urban migration in south-east Brazil

Often fathers and older sons move first and the rest of the family follows later.

Rural to urban migration is common in many LEDCs. Many rural people in Brazil migrate to the cities in the south-east of the country because:

- high population increase causes land and food shortages
- people believe the urban area is a place of 'bright lights' and opportunity
- crop failures, lack of capital, absentee landlords and lack of education mean a life of rural poverty
- farm mechanisation causes unemployment.

Case study of regional migration in Italy

Note the way the push-pull model helps you to understand this migration.

The major movement is from the south, the Mezzogiorno, to the north of Italy. The main factors are:

- low standards of living based upon agricultural employment
- lack of educational opportunities
- steep slopes and poor communications make the movement of goods difficult
- lack of raw materials and poor power supplies
- poor agricultural development because of poor soils and lack of water in summer
- underdeveloped tourist industry so lack of jobs
- failed attempts by government to introduce new industry to the area
- attractions in the north include large cities, rich industry, water supply and better conditions for farmers.

Case study of economic migration in California

Mexican workers started to arrive in the 1950s across the border into California. These Spanish-speaking people, called Hispanics, moved into California to improve their standard of living, find jobs, better education and health care. Today the men move first and then bring their families in once they are established. Many of the migrants are illegal but they are willing to take the harder, dirtier, seasonal and low paid jobs in agriculture, the building industry, hotels and restaurants. Many find themselves living in very poor housing but they are still better off than staying in Mexico.

Population problems and issues

MEDCs and LEDCs face problems arising from population change. They need to find solutions to these problems that encourage sustainable development.

The social, environmental and political problems of population growth in LEDCs

Many LEDCs have high rates of population growth because of high birth rates and declining death rates. The problems caused by this growth include:

- unemployment
- housing shortages
- lack of food
- shanty towns
- crime
- prostitution
- strain on services
- overcrowded buses
- poor sanitation
- poor medical care.

However, this growth provides a labour force and an increased domestic market.

Table 8.3 Advantages and disadvantages of population change

	Areas losing population	Areas gaining population
ADVANTAGES	• lowers population pressure • reduces unemployment • fewer mouths to feed • birthrate falls • money received from migrants • farms become bigger	• increases working population • migrants willing to work in unpleasant jobs • migrants work for lower wages • brings new culture to an area • some may be very skilled • ready to develop informal economy
DISADVANTAGES	• young male workers leave • farm production falls • young, educated people leave • elderly population left behind • services deteriorate	• dangers of racial tension • may put local people out of work • services become overcrowded • may be poorly educated • may not be able to afford local housing • may build shanty towns • may bring disease

The problems of ageing populations in the EU

A growing problem needing special provision; good for exam answers.

Life expectancy is increasing in many countries because of better diets, housing and healthcare as well as less physically demanding jobs, e.g. the 'greying' population in Sweden and the UK. Birth rates have also fallen in these countries. The problems caused by ageing populations include:

- increased funds needed to meet the demand for health services
- specially designed houses and serviced blocks of flats, e.g. hand rails, non-slip floor, emergency buttons
- provision of leisure activities for the elderly
- increased dependency and burden on the working population
- shortages of labour and the need to bring in workers such as nurses and doctors from other countries.

However, older people do undertake many worthwhile tasks, many of them voluntary, as well as supporting their grown-up children as they raise their young families.

The problems of rural to urban migration in LEDCs

Know at least one example of source and destination.

Migrants arrive in cities with little money and few skills for urban jobs. They cannot afford housing and are forced to settle in temporary dwellings made from scrap materials such as corrugated iron, wood, cardboard and cloth. These residential areas have:

- little or no sanitation and poor water supply
- unsafe electricity taken illegally from the National Grid
- few health facilities and widespread disease
- overcrowding, unemployment and crime.

The effect of this movement is that rural areas do not have enough labour; women with the help of the very young and the elderly often run small farms. As a result, farming output falls.

The problems of urban to rural migration in MEDCs

With improvements in public transport and increased car ownership, people are moving away from large urban areas to live in the smaller market towns and villages nearby. This causes problems for these country areas including:

- population increase places a strain on housing and health services
- increase in traffic creates congestion, noise and pollution
- schools may have to be expanded
- housing prices rise and local young people are not able to afford to buy in the area
- loss of open space as developers build houses to meet the demand
- public transport becomes crowded as commuting increases.

Know the effects on source and destination areas.

The effects on the urban area are not always positive, as the people left behind are often those with fewer skills and some social problems. The arrival of affluent people in a village can revitalise the area allowing shops to remain open, the village school to prosper and clubs and societies to flourish.

The problems and issues raised by refugees

Some people are forced to leave their home area or country. They are called **refugees** and estimates suggest there are 30 million of them in the world. The reasons people become refugees include:

- persecution due to race, religion, nationality, social group or political beliefs, e.g. the Vietnamese 'boat people' in 1975 with the fall of South Vietnam; and the Muslims in Serbia due to the ethnic cleansing by the government
- civil war between rival ethnic groups such as that in the 1990s in the African country of Rwanda
- natural disasters such as earthquakes, volcanic eruptions, floods and droughts, e.g. Ethiopians leaving the Sahel region due to drought at a time of high population growth and poor harvests.

The problems and issues raised by these refugees include:

- they arrive in neighbouring countries that do not have the resources to cope with them, e.g. Ethiopian refugees filled camps in nearby Somalia which is another very poor country
- they usually arrive in large numbers in a very short time and have to be housed in tented camps with poor sanitary and water facilities; disease often breaks out
- they need food, water, medicine and shelter.

After the emergency some refugees return home. Others stay in their new country. They may bring valuable skills and expertise and find good employment. Others fill menial jobs that local people will not do. However, there are refugees who do not settle; they find the language, currency and customs of their new country difficult. For a sense of security they tend to stay together in the same neighbourhood and create a small piece of their own country. Racial conflict can occur.

Fig 8.9 Some examples of international migrations

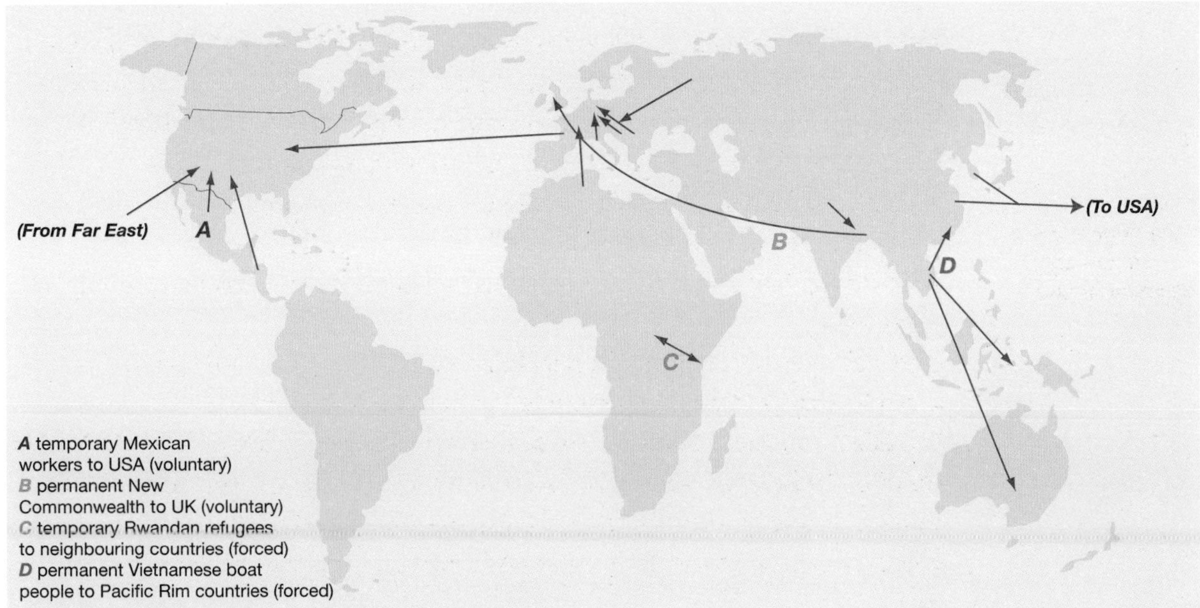

Table 8.4 The reasons for international migration

FOR	AGAINST
Spiritual reasons	
• search for greater enlightenment • to gain greater understanding of life • persecution by authorities	• compromise spiritual values in order to survive • challenges to personal views by people in the new place • tolerance of views held
Moral reasons	
• to get a job so that the family can be supported • responsibility to solve problems of 'home area'	• danger of being unemployed and forced into crime or prostitution • worries that 'new' area may not be what it seems
Ethical reasons	
• duty to earn and send money back to 'home' area • attitudes of local authorities to problems in 'home' area	• fear of failure • concerns over the way migrants are treated in 'new area'
Social reasons	
• to obtain a better education for the individual or family • to access a wider variety of entertainment and social experiences	• break up of the wider family, split with parents • loss of friends and decline in community left behind
Cultural reasons	
• to experience a different and maybe richer cultural scene • opportunity to express themselves more openly	• neglect of personal heritage and culture • dangers of entering oppressive regime

Strategies for coping with population change

Strategies for reducing population growth in LEDCs

Many LEDCs face increasing population totals and more mouths to feed. This takes resources away from plans to improve the **quality of life** of people. The strategies adopted by LEDCs include:

- **birth control** using contraceptives, abortion and sterilisation (tried in India with limited success)
- raising the level of education, especially for girls
- encouraging delayed marriages or raising the age at which people may marry
- making polygamy (multiple spouses) illegal

Many candidates remember the 'one-child policy' but say it comes from Japan!

- raising the status of women to have a view on their number of children
- better health removes the need to have many children in case some die
- financial benefits to couples with small families
- providing employment opportunities and careers especially for women
- China has a **one-child policy** that encourages couples to have only one child. This policy includes:
  - birth control and fines if more than one child is born
  - increased taxes for couples with more than one child
  - public education and advertisements on the reasons for the policy
  - salary bonuses of 10% for those with one child
  - priority in education, housing and health for one-child families.

This harsh policy has been relaxed particularly in rural areas. It created a generation of 'spoiled, only child'. Some girl babies were abandoned or murdered because parents wanted a son to keep the family name alive.

Population strategy: plans by governments to solve the problems of population growth or decline.

Strategies for increasing population in MEDCs

This has become an important topic so expect examination questions on it.

Some MEDCs have low birth rates, increasing numbers of elderly people, and a falling population total. The population is not at its **replacement level**. These countries face **labour shortages** and do not meet economic plans for improvements in the quality of life of the people living there. The strategies adopted by these MEDCs include:

- in Japan, they have raised the age of retirement to increase their workforce
- in Singapore the government has:
  - encouraged single people to marry
  - encouraged married couples to have more babies
  - provided a Social Development Unit that helps single people to meet each other
  - established child care centres for working women
  - introduced flexible working hours and part-time employment for mothers
  - extended the period of maternity leave
  - given tax rebates to married couples with more than two children
  - subsidised childcare centres
  - given housing priority to couples with more than two children
  - advertised the benefits and joys of marriage on television programmes
  - shown stories of happy family life in newspapers.

Reducing the pressure of population growth

Planned migration

It is worth looking at an atlas to see where these places are found.

Indonesia has planned for the movement of some of its people from the crowded islands of Java (population density 700 persons per km^2) and Bali to the less crowded islands of Sumatra, Kalimantan, Sulawesi and Irian Jaya (population density 3 persons per km^2). This is called a transmigration policy. It is the world's largest resettlement scheme. This has:

- reduced overcrowding and unemployment in Java and Bali
- lowered the pressure on the land in Java and Bali

- eased rural to urban migration
- made use of large areas of forested land on the other islands.

Under the plan, land is given to poor farmers, together with some seeds, a house and food. It has been successful in Sumatra. Some migrants have found jobs on plantations opened on the hilly land. The migrants have more skills than the local people and there have been clashes. The migrants have challenged traditional culture and taken some urban jobs from local people.

Fig 8.10 Transmigration in Indonesia

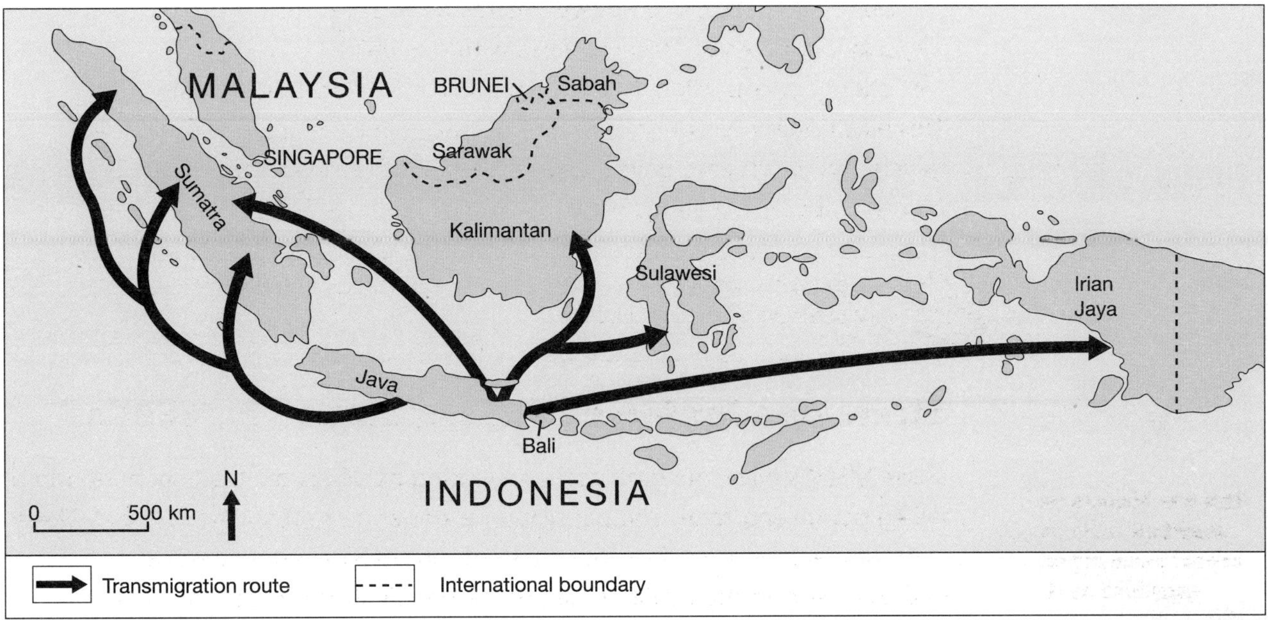

There has been soil erosion in some of the forest areas that were cleared for cultivation. Java remains a very crowded island.

Strategies for coping with ageing populations in MEDCs

There will be purpose-built housing for the elderly in a town near you.

Increased life expectancy means that governments and local authorities in North America, Western Europe and Japan have to make provision for the elderly. The types of provision they need to make include:

- housing which is affordable and easily managed by older people
- health provision to meet the needs of the elderly including specialist cancer and heart units
- car parking and shopping facilities which are suited to people less steady on their feet
- access to social and religious services which will include more home visits
- library and other leisure facilities
- encouraging employers to give jobs to older people, e.g. craftsmen as advisers in DIY stores.

1. Give two advantages for an area losing population through migration.
2. Define life expectancy and explain why it is increasing.
3. Name a group of refugees and explain why they wish to move to another country.

1. See page 119 2. See page 119 3. See pages 120–121

Chapter 9

Settlement

LEARNING SUMMARY

After studying this section you should be able to:

- *describe the site and situation of rural and urban settlements*
- *recognise the functions of settlements and the services they offer*
- *know about urban land use and zones within settlements*

The location of settlements

AQA A AQA B

The location of settlements involves the study of both site and situation of different settlement types.

Different types of settlement

There are two types of settlement, **rural**, e.g. a village, and **urban**, e.g. a city. Residents in rural settlements are mainly farmers or workers from urban settlements. Urban settlements are involved in industry, commerce or administration. Rural settlements are smaller in size, population and population density. Urban settlements include towns, cities, conurbations and megalopolises.

Rural: small number of buildings, agricultural or dormitory function, low density, close community, small number of services, in countryside. Urban: continuous built-up area, with industry and commerce, high population density, acquaintances rather than friends, many services.

The siting and situation of settlements

Site: the features of the place where the settlement is located. Situation: the features of the area in which the settlement functions.

The factors influencing the **site and situation** of settlements include:

- dry point in an area of poor drainage, e.g. Ely, and wet point where water supply is good, e.g. Gretton, Gloucestershire
- spring line, e.g. Princes Risborough
- shelter and defence, e.g. Durham, Edinburgh, Conwy
- resources such as minerals, e.g. coal
- communications: gap towns, e.g. Dorking, route centres, e.g. Crewe, and bridging points, e.g. Worcester – often settlements based on communications became trading centres
- planned: new towns, e.g. Harlow, expanded towns, e.g. Swindon
- resorts, e.g. Bournemouth, spas, e.g. Buxton

Check whether a question asks for a site or situation. If it asks for location it means both.

- ports, e.g. Grimsby, Southampton
- religious centres, e.g. Canterbury
- market centre, e.g. Norwich

These factors were important when the settlement was founded but are less so today.

Fig 9.1 A spring village line

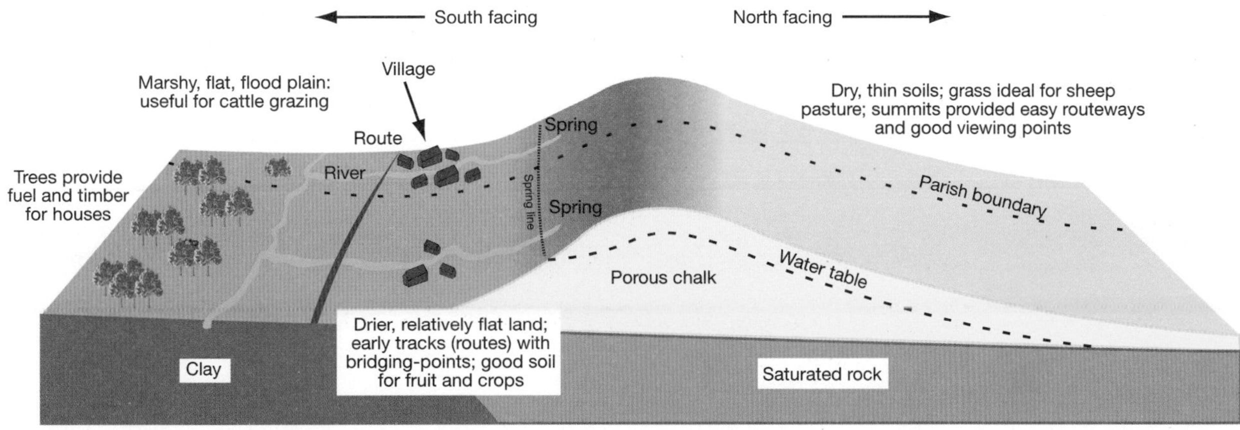

The function of settlements

AQA A AQA C

KEY POINT **Settlement function: describes the main economic activity of the settlement, the jobs of the people working there.**

Settlements can be **classified** according to their main function:

- manufacturing, e.g. Sheffield
- holiday resort, e.g. Blackpool
- educational, e.g. Oxford and Cambridge
- hi-tech industry, e.g. Reading
- pottery, e.g. Stoke on Trent

All larger settlements also act as centres for administration of the area around them.

Know the names and functions of towns in your area.

Services in settlements (includes hierarchy, range of good, threshold, sphere of influence)

Settlements provide housing, administration, industry, commerce and services. The services include local government offices, shopping centres, schools and hospitals. Settlements can be arranged into a hierarchy according to their population size, range of services and distance apart.

Central Place Theory attempts to explain the size and spacing of settlements and the services they offer using the following ideas.

Goods

Goods purchased regularly, such as bread and newspapers, are called low-order or convenience goods. They are usually available in the village or corner shop.

Goods purchased irregularly are called high-order or comparison goods, e.g. television, clothing, furniture. People compare prices on these goods; they want choice and are prepared to travel further to city centres or retail parks.

Range of a good

This is the maximum distance that people are willing to travel for a service. They travel further for high-order, comparison goods.

Threshold population

This is the minimum number of customers needed to maintain a service. Small local shops need a small threshold population buying low-order goods regularly. Comparison goods shops need a larger threshold population because goods are bought infrequently. Very specialist shops, e.g. jewellers, need a very high threshold population and are usually only found in large settlements.

Be able to apply range, threshold and sphere of influence to two settlements you know.

Sphere of influence

This is the area served by the settlement. It is the area from which people travel to use the settlement or the area to which bus services run and deliveries are made. Newsagents have small spheres of influence, secondary schools have larger spheres than primary schools and hospitals have very large spheres of influence.

Hierarchy

The hierarchy of settlements extends from metropolitan areas, through cities, towns and villages to hamlets. The number of services offered and the sphere of influence increase up this hierarchy.

Fig 9.4a Hierarchy of settlements

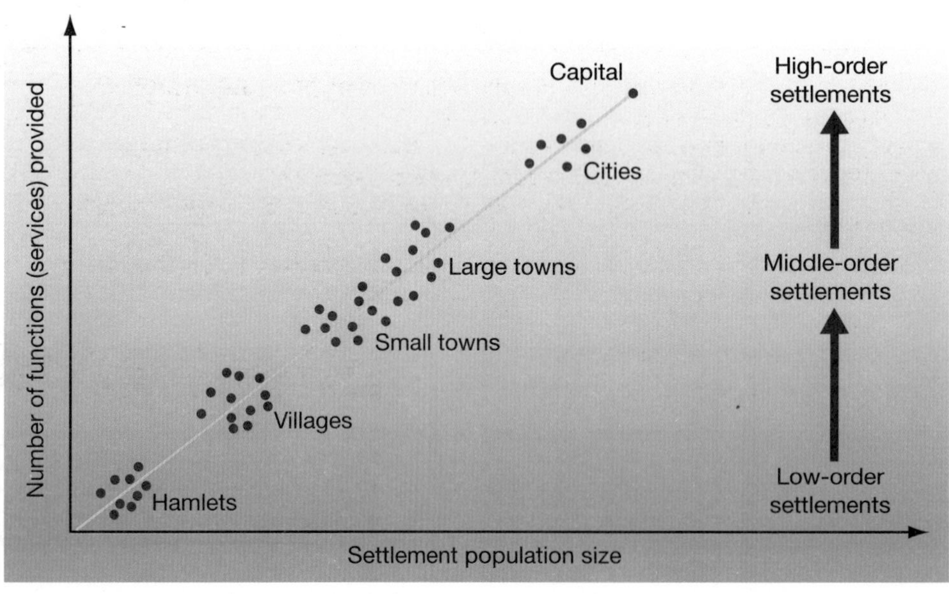

Fig 9.4a shows there is a large number of low-order settlements and a small number of high order settlements.

Fig 9.4b Hierarchy of shops and services

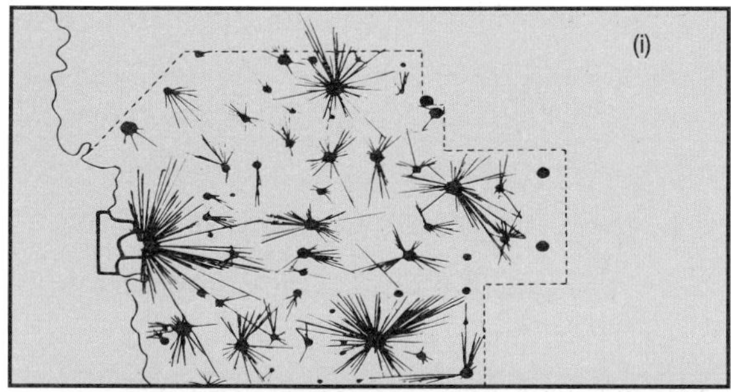

Fig 9.4b shows the pattern of movement of people to settlements for (i) food, (ii) solicitors and (iii) hospitals.

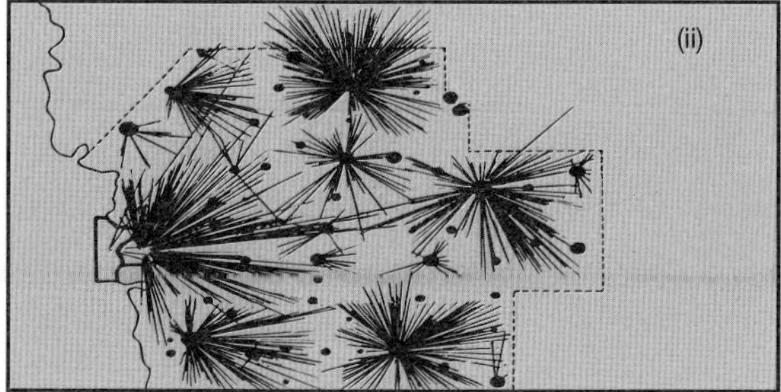

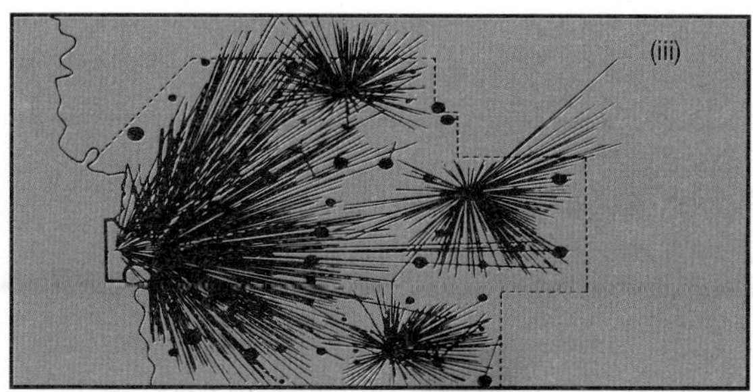

Fig 9.4c Sphere of influence

Fig 9.4c shows the spheres of influence of a town for three different services.

PROGRESS CHECK

1. What is the difference between the site and situation of a settlement?
2. Name an example of each of the following:
   - a wet point settlement
   - a resort
   - a new town
   - a market town.
3. Which of the following has the largest sphere of influence:
   - a newsagent
   - a supermarket
   - a furniture shop
   - a public house?

1. See page 124 2. See page 124–125 3. See pages 125–127

Models of land use in settlements

AQA A AQA B

KEY POINT

Surveys of land use in settlements show areas with similar land use. These are called land use zones. For example, a zone of lower-quality terraced housing can be found near to the CBD (Central Business District).

To simplify and explain the complex pattern of land use in settlements a number of models have been developed. One of the three main models is the **concentric zone model**.

No settlement fits exactly to any model. Hills, rivers and historic buildings such as a cathedral, castle or large house may distort the concentric rings. Matching the model to a settlement helps increase understanding of the special features of a place.

Fig 9.5 The concentric zone model

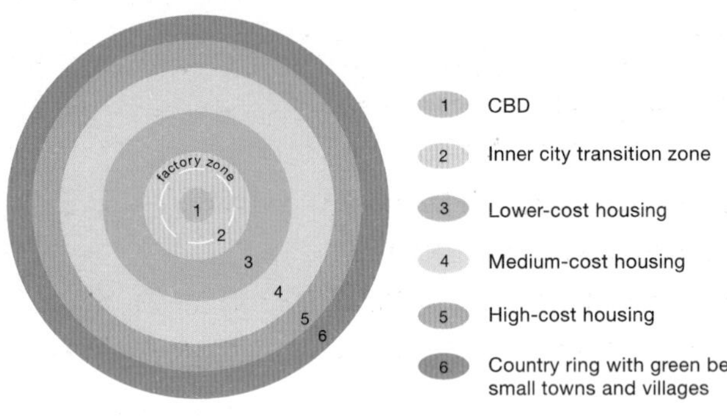

The concentric model

This model was based upon studies of American cities and particularly on Chicago in the 1920s. It has the Central Business District (CBD) at its centre. This zone is mainly made up of shops and offices. Around the CBD is the zone of transition or twilight zone or inner city. This is an area of older housing, declining industry and some derelict land. Outside this are three housing zones, which increase in quality away from the centre.

Land use in the Central Business District

Think of ways to find the edge of the CBD.

Competition for land makes land prices high in the CBD. The CBD contains the main retail and commercial premises, covered shopping arcades, pedestrianised areas, major public buildings and administrative headquarters. Shops sell high-order goods and high-rise buildings are present. There is a high daytime population but few people live there. Land use and buildings are continually changing in the CBD. The congested CBD faces competition from out-of-town centres. This area is called '**downtown**' in the United States.

Land use in the inner city

Census data for urban wards can help to identify the inner city.

The land around the CBD is called '**the inner city**', the 'twilight zone' or 'zone of transition'. In this zone there are derelict old factories and poor terraced housing, built in the nineteenth century. Some of these 'brownfield' sites are being replaced by CBD buildings, large warehouses or new housing, roads and railways. During the period of redevelopment they become temporary car parks. Many people living in the inner city suffer deprivation (unemployment, overcrowding, crime, and pollution). Government policies have tried to redevelop, renew, rehabilitate and gentrify these problem areas, e.g. Hulme in Manchester, Toxteth in Liverpool.

Greenfield sites are open land that has not been built on before.
Brownfield sites are vacant or under-used land that has been built on and is ready for redevelopment.

Land use in the suburbs

Housing is expensive in the suburbs but much of it was built over 40 years ago and is now deteriorating.

Away from the CBD, land prices fall. Houses and gardens become bigger and people living there can use their car or public transport to get to work or the city centre. As population and mobility increased, cities sprawled over the surrounding countryside. Factories and offices have moved from the inner urban area to 'greenfield' sites in the **suburbs** near to motorways or ring roads. Many local authorities moved people from the inner areas to large council estates in the suburbs. Some of these have become problem areas. People living in the suburbs tend to segregate into areas of similar class, income or ethnic background.

The country ring

Counter-urbanisation is the movement of people and economic activity out of larger settlements into the surrounding countryside and towns.

Land values are lower in these areas. Since the 1970s, housing, commerce and industry have moved out of the urban area to the settlements in the countryside. This is called **counter-urbanisation**. Quality of life is better in these areas; there is more space and less congestion, crime, vandalism and pollution.

Land use in settlements in LEDCs

The concentric zone and sector models were developed from studies of settlements in North America. They have been revised for cities in LEDCs.

- The function of the CBD remains the same as in the MEDC city. However, much of the area will be redeveloped with modern high-rise blocks, shopping malls, hotels and apartments. The area may contain one or two older buildings in an unusual style that are relicts of a colonial past.
- The inner zone contains middle- and high-class housing. There will also be blocks of flats for the working class.
- The outer zone includes large houses and bungalows for the rich as well as squatter settlements.
- Industry is to be found along transport routes into or around the urban area.

Fig 9.6 Model of land use in an LEDC urban area

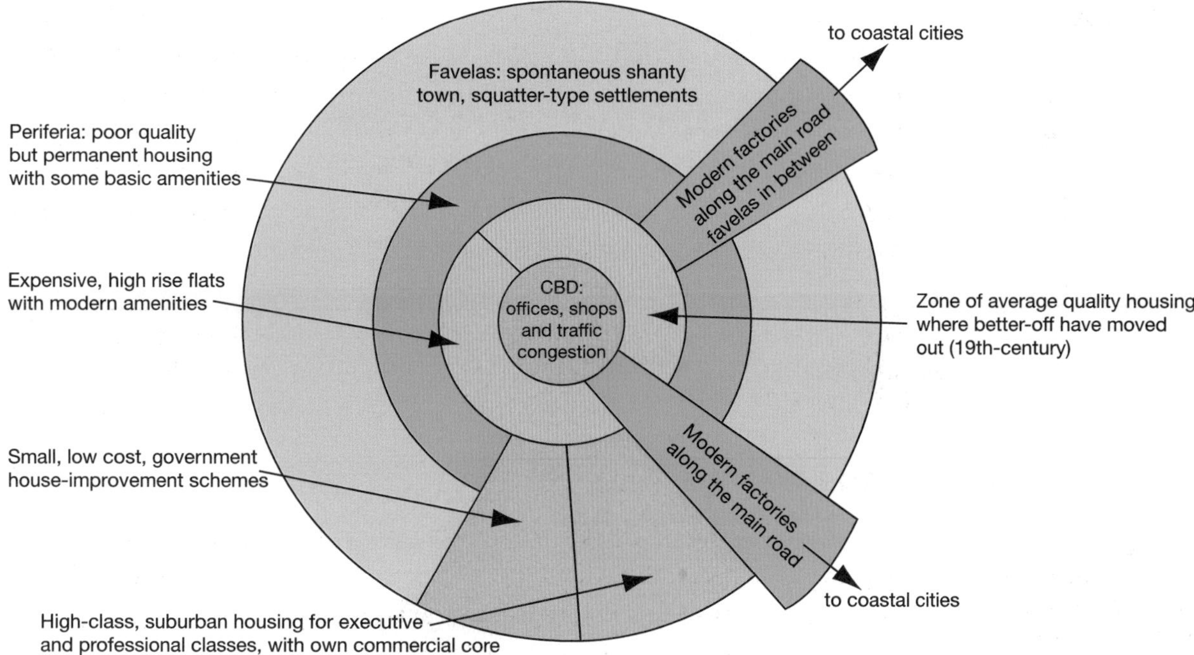

Shanty towns are a special feature of large settlements in LEDCs. Shanty towns tend to house the poor and other poor people arriving in the settlement from rural areas. The residents are squatters with no legal rights. Housing is primitive, sometimes only one room in which a family eats and sleeps. Few of the houses have toilets or running water. Rubbish collection is unreliable, and disease and crime are frequently found there.

Factors affecting land use in settlements

AQA A AQA C

What are the factors influencing the arrangements of **land use zones** in a settlement? An answer to this question allows planners to rearrange the land use in an attempt to improve the quality of life of the people living there.

The factors influencing land use in settlements include:

- accessibility
- competition for land
- the mixing of land uses
- the specialised needs of certain land uses.

Remember examples of these factors and mention that they can change, e.g. London Docklands became redundant as the size of ships increased but has been revived as a financial services and newspaper printing area.

Access

Different types of transport make some places in the settlement more accessible than others. For example, bus and trams routes allowed the countryside to become accessible because people did not have to walk to work any more and the suburbs developed. Shops are located in very accessible places. This means the CBD, at ground floor level, is dominated by shops. However, as congestion has increased, larger shops have moved to the edges of settlements to retain their accessibility and provide space for car travellers to park.

Another example of **accessibility** was that housing used to be built next to factories. This still happens in some countries, e.g. Singapore. In the UK, industry has moved to more accessible locations on ring roads and motorways where they are close to settlements. People travel to these industrial estates in their cars.

Competition for land

Competition for land is at its highest in the CBD because of the good accessibility. Access is high because the centre is the focus of rail and bus routes. As a result, land prices rise. **High rents** mean only national chain stores, financial services, hotels and offices can afford the rents. Oxford Street and Regent Street in London and Fifth Avenue in New York are good examples.

A sign of the competition for each piece of land is the presence of **high-rise office blocks**. This means each piece of land can be used several times.

In some CBDs, the competition for land has fallen as out-of-town centres have become more accessible and attractive. This has caused a decline of the CBD; shops have become empty or occupied by temporary bargain stores or charity shops.

Mixing of land uses

Some land uses do not mix. It is unlikely that high-income residential areas will be found close to heavy industry or airports. On the other hand, industries that need a large labour force, such as the clothing industry which requires large amounts of female labour, tend to locate near to large housing estates.

Pressure on land means that some new developments do mix unsuitable land uses. For example, new housing is build alongside new bypass roads but the houses have double-glazing and high fences to deflect the noise.

Specialised needs

Good examples of this include:

- shipyards on deep-water estuaries
- airports on extensive areas of flatter land
- warehouses and distribution depots near to motorway interchanges.

PROGRESS CHECK

1. No settlements ever match an urban land use model perfectly. Why is it useful to study these urban models?
2. How can census data for small areas in a settlement be helpful to town planners?
3. Why is housing so expensive in the suburbs when land values are lower there?

1. Matching the model suggests a general explanation. Variations from the model suggest local factors. 2. Shows areas of deprivation and other groups with special needs. Also assesses housing quality. 3. Tends to be on larger plots of land with bigger houses. Demand is high, especially from families with several wage-earners.

Chapter 10 Urbanisation

After studying this section, you should be able to:

- ***know the causes of urban growth***
- ***understand the problems of urban areas***
- ***recognise some planned solutions to urban problems***

Urban growth

In the last 150 years, urban areas have grown rapidly in population and area through **natural increase** and the **migration** of people from the countryside. Some recent examples are shown in Table 10.1.

Table 10.1

City	Population 1970 (millions)	Population 2005 (millions)
Mexico City	8.6	24.0
Sao Paulo	7.1	20.0
Calcutta	7.0	13.0

Fig 10.1 Projected urban population growth

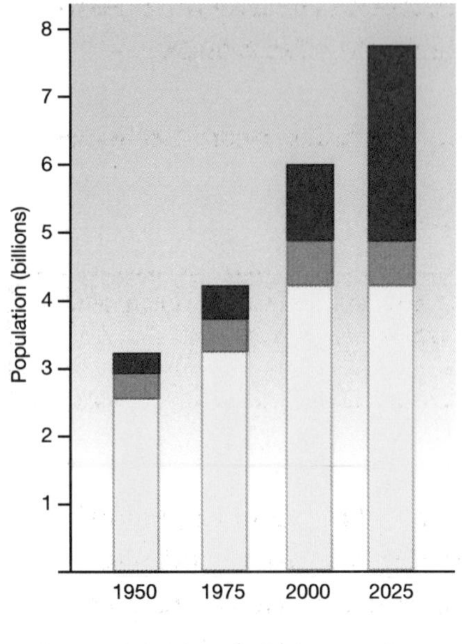

Mexico City occupied 50 km^2 in 1950 and grew to 1100 km^2 by 2000. Urban growth is taking place throughout the world. In 1950 the world's urban population was thought to be one billion but by 2000 it was estimated to be over four billion. Most of the increase today is taking place in LEDCs.

In 1900 only London and Paris had populations greater than one million (**millionaire cities**). By 2000 there were over 300 of these. Cities with more than eight million people were called **megacities**. In 1950 there were only two megacities: New York and London. By 1990 there were 21 megacities and 16 of them were in LEDCs. Recently the definition of a megacity has been changed to a city with a population of ten million, and in 1995 there were 15 megacities, with 11 of them in LEDCs. At current rates of growth there are expected to be 33 megacities by the year 2025, with 27 of them in LEDCs.

One billion is one thousand million (1 000 000 000).

Fig 10.2 The world's megacities

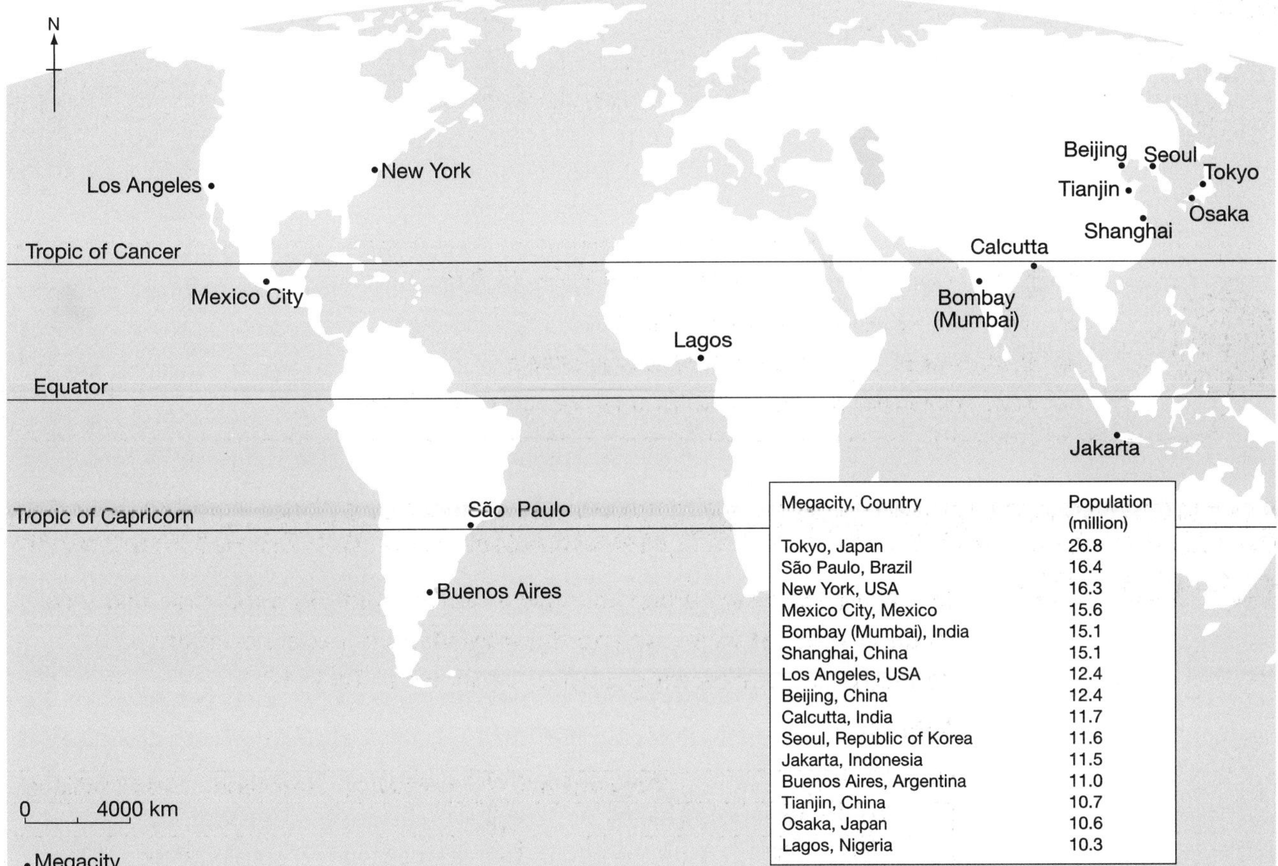

Megacity, Country	Population (million)
Tokyo, Japan	26.8
São Paulo, Brazil	16.4
New York, USA	16.3
Mexico City, Mexico	15.6
Bombay (Mumbai), India	15.1
Shanghai, China	15.1
Los Angeles, USA	12.4
Beijing, China	12.4
Calcutta, India	11.7
Seoul, Republic of Korea	11.6
Jakarta, Indonesia	11.5
Buenos Aires, Argentina	11.0
Tianjin, China	10.7
Osaka, Japan	10.6
Lagos, Nigeria	10.3

Finding the total population of a city is difficult because it keeps changing. People come and go and its boundary moves.

Some countries, especially LEDCs, have one city that is much larger than all the others in the country. It is called a **primate city**. Such cities were usually established as the capital in colonial times. Primate cities include Lima (Peru) and Paris. The reasons for the development of primate cities include:

- high rural to urban migration
- the development of a core region containing the main economic activities
- the capital city of a centralised government
- when capital and major port are in the same city.

The words conurbation or megalopolis are used to describe an urban area where a city has grown to absorb several larger towns, e.g. the Manchester conurbation.

The causes of urban growth

The three main causes of urbanisation are rural to urban migration, natural increase in the urban population and the redrawing of urban boundaries.

- Rural to urban migration (see page 117) occurred in MEDCs in the past as people moved to the new factories, ports and mines found in the urban area. Towns rapidly expanded to become cities. In LEDCs it is occurring today as people look for jobs, education and healthcare. Poor harvests, lack of agricultural land and farm mechanisation have pushed them away from the rural areas.

You need to be able to describe suburbanisation, gentrification, green belts and counter-urbanisation. Know why some people suggest that suburbanisation has helped cause the decline of inner city areas.

- Natural increase is taking place as healthcare becomes more widely available and medical advances combine with a better quality of life to increase life expectancy.
- Urban boundaries are redrawn as the urban areas expand to house all the people living there.

In the UK, cities spread as public transport improved. Suburbs of semi-detached houses developed on new housing estates. Light industry provided nearby employment or the new residents could **commute** to the CBD by bus and train. In the 1980s, business parks and out-of-town shopping centres encouraged this trend. However, since the 1950s, many local authorities have established green belts around the edge of the city to stop it sprawling any further across the countryside.

In the 1990s, better-off people, particularly families and the recently retired, moved out of the urban area to live in smaller settlements in the nearby countryside. They were moving away from the congestion, pollution, crime and noise of urban life. They found the rural environment more attractive and could travel quickly into the urban area if they needed to. This movement out of the urban area, which was first noticed in the USA in the 1970s, is called '**counter-urbanisation**'. Unfortunately this outward movement leaves behind lower income groups, including recent in-migrants from ethnic minority groups and the unemployed. Now some cities are attracting young people back by imaginative **redevelopment** of old factory and warehouse sites, e.g. Manchester and Leeds in the UK and Amsterdam in the Netherlands.

Whilst cities in MEDCs have tended to stop growing, those in LEDCs continue to grow and spread.

Urbanisation is the process leading to an increased proportion of the people in urban areas.

Urbanisation

Urbanisation has been taking place for 200 years but there has been a dramatic increase in the last 50 years. It began in MEDCs with the Industrial Revolution; today between 70% and 80% of the world's population lives in urban areas (Table 10.2).

Table 10.2

Proportion of world's population living in urban areas	
1950	29%
1990	46%

Be clear about the difference between urban growth (increase in area and population) and urbanisation (increased proportion living in urban areas).

In MEDCs the rate of urbanisation is slow because the majority of the population already live in urban areas. In addition, rural areas are more attractive and rates of natural increase are lower. Rates of urbanisation are much higher in LEDCs.

Urbanisation brings advantages and disadvantages. On the one hand more facilities such as education, healthcare and transport may develop. But disadvantages include many problems with inadequate services and lack of housing and jobs.

Problems caused by rapid urban growth

AQA A  AQA C

KEY POINT **Rapid urban growth of population and city size causes a whole range of problems.**

Housing

This is a common problem. Housing is often of poor quality, insufficient or too expensive. Where there is not enough housing, prices rise and the poor are forced to live in very low-quality houses. In some cities, poor people build substandard shelters in **squatter settlements** on the edges of urban areas. (These are called *kampungs* in Jakarta, Indonesia.) They use plants, corrugated zinc sheets, and other scrap materials. Because of the shortage of land, these camps are often built on flood plains where there is a danger of flooding, on steep hillsides where mudslides occur or close to railways and airports where noise makes life difficult.

These **squatter settlements** have many problems including:

- very poor-quality housing
- poor or non-existent water supply
- unreliable electricity often taken illegally from the mains
- little or no sanitation
- poor drainage for rain and household waste water
- little or no waste collection service and rubbish dumped on open spaces
- overcrowding and high risk of disease, e.g. tuberculosis
- little employment because the factories are too far away.

The lack of affordable housing also causes homelessness. At its worst these people live and sleep in the streets. They cover themselves at night with cardboard and old blankets. On cold wet nights they sleep in office doorways and hope to get some heat as it escapes from inside the building.

Transport

Many urban areas have serious transport problems. Bangkok is said to have the worst **traffic congestion** in the world. When traffic congestion occurs, average speeds fall and results in frustration, increasing costs, people being late for work or appointments and falling productivity. This is mainly the product of:

- the growth of vehicle ownership in recent years, e.g. 15% annual increase in Indonesia and especially Jakarta
- poor road layouts especially in rapidly growing LEDCs
- poor public transport with buses and railways unable to keep up with the demand, e.g. trains in Tokyo (Japan) and London and buses in Calcutta (India) and Harare (Zimbabwe) are packed in the rush hours
- the rolling stock of public transport, e.g. buses and trains, is worn out and breaks down frequently
- breakdowns and accidents cause additional delays and congestion as the network reaches capacity.

The M25 motorway around London is sometimes described as the largest car park in the world!

Unemployment

Many of the people migrating to cities are male. They come to find a job and then ask their families to join them. Many do not have the skills for urban jobs and find themselves unemployed. They join the informal economy and run their own small business providing goods and services. In Beijing, China, they may run their delivery business on the back of a bicycle and their office would be the pavement. Cheap mobile telephones help.

Pollution

Urban areas suffer problems of **air, land, water and noise pollution**.

The causes of air pollution are:

- **industrial activity**
- **increased numbers of vehicles on roads**.

These put dust particles and higher levels of dangerous gases, such as carbon monoxide and sulphur dioxide, into the air. These gases can damage the health of people in the area especially the elderly, the asthmatic, those with bronchitis and eye problems. Air pollution also increases the dangers of fog and traffic accidents. Los Angeles and Mexico City have become famous for their high levels of air pollution.

Make a two-column table and summarise these problems and their planned solutions.

Land pollution occurs when domestic and industrial waste is left uncollected. For example, domestic rubbish is usually collected once a week and its accumulation can cause pollution. In squatter settlements there may be no proper collection of domestic rubbish. Some industrial areas have no safe way of disposing of dangerous wastes and dump it on open spaces. Official and unofficial landfill sites bring the problem of flies and diseases such as diarrhoea and malaria.

Water pollution can be a major problem in urban areas. Rivers and canals can become polluted by domestic and industrial waste. In some LEDCs only 40% of houses are connected to the sewerage system. Even in some MEDCs much waste material is dumped through long pipes into the sea. This can threaten wildlife and the safe use of tourist beaches.

Noise pollution comes from industry, construction, and traffic including trains, motor vehicles and aircraft. Noise pollution adds to the stress of living in crowded urban areas.

Water supply

Some urban areas have inadequate water supplies. This is especially true when rainfall is low and the settlement relies on groundwater supplies. In addition, large quantities of water are lost through leaking pipes. The rapid growth of population and industry increases the demand. In richer areas modern domestic appliances and swimming pools also increase demand. It is estimated that water demand increases at three times the rate of population increase.

In cities in LEDCs many people do not have piped water to their houses. They may have it delivered in expensive water tenders by private water sellers or use a communal tap (one tap may serve 500 people and only work for a few hours each day). In the worst cases people use polluted wells and open pipes. For example, in some Indian cities clothes are washed in the street from storm water drainage pipes.

Water can also be of poor quality when water treatment plants cannot cope with demand. Poor water quality brings health threats. In the UK people are buying more bottled water, as they dislike the chemical taste of tap water.

Solutions to the problems of urban areas

AQA A AQA B AQA C

KEY POINT **Urban areas have many problems that local and national governments attempt to solve through town and country planning.**

Housing

There are housing problems in both MEDCs and LEDCs. The solutions in MEDCs have similarities and differences from those in LEDCs. The main **solutions to the housing problems** in the urban areas of MEDCs are:

- use brownfield sites such as empty houses, former factories, office blocks, flats above shops or railway sidings. This land is often in poor condition and expensive to restore. However it has many advantages including the facilities in the town within walking distance, existing infrastructure and saving green fields
- the extension of existing towns by bolting on small suburban estates, often within the line of a newly built bypass
- building a small number of houses in several villages by infilling or small developments
- building new settlements such as new towns where all the infrastructure is new and houses can be built to the latest standards
- selecting one key village to be developed to the size of a small town with additional schools, bus routes and all the other services provided.

When a local authority plans to develop housing in an area it consults the people living there and asks for their suggestions. These schemes for housing estates in villages and rural areas usually attract opposition from local people. They are called NIMBYs by some people (Not In My Back Yard). The reasons given for opposition include:

- lack of employment locally
- increase in cars travelling on narrow village roads
- children travelling to school will cause congestion twice per day
- the small amount of public transport will not be able to cope
- the houses will spoil the views and newcomers will not fit into the community.

In LEDCs the solutions are different and include:

- discouraging rural to urban migration by improving conditions in rural areas by developing growth poles with jobs, housing, schools and healthcare. In Indonesia, rural people have to get a permit before they are allowed to move to an urban area
- improving living conditions in squatter settlements by building paved roads and pavements, laying drainage systems, improving water and electricity supply, building schools, clinics and community facilities. Unfortunately improvements attract even more people

- building subsidised public housing which may be low-rise, medium-rise walk up, e.g. Kuala Lumpur, Malaysia, and high-rise, e.g. Hong Kong and Singapore
- encouraging self-help housing through co-operation between the government, local authorities and the people. Cheap land sites are provided with the basic infrastructure of roads, drainage, water pipes and sewerage systems. People then buy cheap materials to build their own houses under supervision e.g. South Africa, Zimbabwe and Indonesia.

Find a case study of a planned new housing development in your local area and one in an LEDC.

Transport

There is a wide range of solutions to this set of problems:

- **building** new roads and widening existing ones with the building of flyovers, tunnels and bypasses, e.g. Birmingham, Leeds, Kuala Lumpur (Malaysia)
- **maintaining** roads and improving central crash barriers, street lighting and co-ordinated traffic lights
- **controlling** the increase in vehicle numbers, especially cars since each carries so few people, by increasing the costs to motorist through taxes on petrol, road taxes, tolls on busy roads (these actions are very unpopular with voters)
- **encouraging** the use of public transport that makes a more efficient use of the available road space. This can be done by subsidising fares or building new and faster transit systems, e.g. the MRT in Singapore, the MTR in Hong Kong, the new tube lines in London, tram systems in Manchester and Sheffield and BART (Bay Area Rapid Transit) in San Francisco
- **discouraging** unnecessary travel by making it more expensive, e.g. high costs of parking, taxes to enter restricted zones (Singapore), and encouraging the use of faxes, telephones and electronic mail.

Some countries have introduced traffic lights that show the number of seconds to go to the next change of the lights.

Employment

Tourism is one of the keys to increased employment. Tourists bring much needed foreign currency and demand many services. Many city authorities are allowing the informal sector to grow. Jobs include shoe shiners, street vendors, newspaper sellers, guides, food and drink sellers, taxi drivers (cycles/motor cycles), and makers of pottery and crafts, soaps and ornaments.

Pollution

The main types of solution that governments can use are:

- **fines** for breakers of environmental laws, e.g. if factories or cars emit more than the allowed level of gases or people litter the streets or discharge chemicals into water sources
- **public education** about the causes and consequences of pollution and encouraging people to be more thoughtful, e.g. limited hours for construction working called 'making pollution an individual matter'
- improving **waste disposal**, e.g. better sewerage pipes, twice-daily refuse collection in busy cities, provision of plastic sacks and 'wheelie bins'.

Water supply

The main solutions are:

- agreeing contracts for supply from areas with plenty of water
- building more reservoirs to store water
- investing in desalination plants to convert seawater into fresh water (but this is very costly)
- stopping water pollution at source
- building more water treatment plants
- installing piped water into older housing and replacing lead and steel pipes
- conserving water through education programmes
- saving water by its re-use, e.g. using clothes washing water to flush toilets
- improved industrial technology to recycle water.

In addition to these plans it helps to reduce the problems if stable governments are in power. They are able to reduce corruption and unrest which in turn attracts foreign investors who are attracted by the low wages and lack of trade union activity. Stable governments are also able to obtain foreign aid to repair the infrastructure and provide skills training for the local people. Some city governments have found that when they involve local people in decision-making their schemes are more successful.

Urban change in the UK: problems and planned solutions

AQA A AQA B AQA C

KEY POINT **In the UK there are very strong laws to provide solutions to urban problems.**

The regeneration of cities

In the 1960s, the areas around the centres of cities in the UK had many problems. They had become run-down and overcrowded. Housing, built in the nineteenth century, was very poor and cramped with people sharing damp living areas and few had central heating. In these areas factories had closed due to poor access and the need for more space; the factories that were left were old-fashioned and unemployment was growing. The better-off and more skilled people moved to the suburbs, often following industrial employment. The people left in the inner city suffered from multiple deprivations.

The census of population offers measures of deprivation including overcrowding, single parent families, people living alone, unemployment, lack of basic housing facilities and no car.

The first round of plans to solve the problems, called comprehensive redevelopment, involved urban **redevelopment or renewal**. Slum clearance took place and many terraced houses were demolished. Clearance spread beyond the inner city to include the older suburbs. The people living in these areas were re-housed in new estates on the outskirts or in tower blocks in the inner city. Sheffield with the Hyde Park and Park Hill redevelopments became famous for this policy.

However, these new inner city estates became problem areas as the blocks were poorly built, communities were disrupted and crime grew. The tower blocks did not have any sense of community, they were noisy, insecure with dangerous dark corners in stairwells and the lifts and communal central heating frequently did not work.

Later schemes **rehabilitated** the older properties with damp-proofing, central heating, bathrooms, new doors and window frames, cleared spaces for parking and made open space into parks. Some of the tower blocks were demolished and low-rise blocks turned into houses with gardens.

Several governments have introduced a range of schemes to help the people still living in the **inner city**. These schemes include the views of local people and have included more provision for job creation and community activities.

A good case study is Hulme in Manchester where the blocks of the 'Crescents' have been replaced by low-rise family housing. People are now moving to live in Hulme.

Governments have noted that there are large areas of unused and derelict land in cities (**brownfield land**). Developers are encouraged to redevelop this land rather than use greenfield sites.

One famous scheme is the **London Docklands**. This government scheme regenerated the area back to effective use and included:

- reclaiming derelict land
- building new roads, the Docklands Light Railway and London City Airport
- new water, gas and electricity services
- building office blocks such as Canary Wharf
- creating many new jobs
- building new homes and refurbishing older ones to increase the resident population
- constructing new shopping facilities and a technology college
- planting over 100 000 trees.

The scheme has been criticised as failing to meet the needs of the local people. The homes provided were very expensive and the jobs were in high-tech industries and unsuited to local low-skilled people.

In the 1990s some run-down areas of cities have become fashionable places to live. The old houses have been bought cheaply and renovated by a wealthier middle class. This is called gentrification. The areas chosen usually have good access to office jobs in the CBD.

Green belts and new towns

Green belts

In the 1930s town planners tried to keep the countryside close to the people living in London by preventing further urban expansion. They established a green belt around the urban area and made planning permission within it very difficult to obtain. In the 1950s, nearly all UK cities established green belts with the added idea of providing space for recreation.

New towns

New towns were built in the 1950s around London, Glasgow and Birmingham:

- to help prevent urban sprawl
- to act as growth poles to draw industry away from the congested cities
- to relieve overcrowding and create better housing within the urban area
- to provide housing and jobs for people willing to move from the city slums.

Over **30 new towns** have been built including Harlow, Stevenage and Cumbernauld. Some new towns in the UK have been successful and this planning idea has spread all over the world.

In some places the government has **expanded an existing town** instead of building a new one. This has been successful in Swindon and Peterborough. Critics of green belts say that, although they preserved a green area around the urban area, development of housing and factories 'leapfrogged' them and moved into the countryside beyond. Great pressure to be allowed to build on the green belt continues, especially in the south east of England.

Government planning in LEDCs

A number of LEDC governments have planned schemes to ease the problems they have in cities. These include:

- **Brasilia**, the new capital city of Brazil completed in the 1960s, to act as a growth pole to pull development away from the cities of Rio de Janeiro and São Paulo in the south east of the country. Whilst the city has had some successes, there is still heavy pressure on the older cities
- **new towns** have been built in Singapore and Egypt to divert growth from Singapore City and Cairo. They have been successful in re-housing people from the overcrowded areas in the main city
- **growth poles** have been established in rural areas to reduce the push factors driving people into the cities. This strategy has been used in Brazil and Zimbabwe.

The scale of urban growth in many LEDCs is so great that the city authorities have had to look for new ways to solve their problems. More sustainable ways that have been tried include:

- self help schemes where the authorities provide a small plot of land and a small loan to help build a dwelling
- low cost community housing projects including providing sewers, sinks and water tanks
- providing electricity, clean water, schools and rubbish collection
- paving alleys and roads, adding street lighting, public toilets and bus stops
- granting ownership of the land to people 'squatting' in shanty towns and recognising residents' associations.

Despite all these efforts, the problems of large cities in LEDCs continue to grow. Successes are overtaken by more people arriving in the area.

1. Describe the pattern of megacities.
2. Define suburbanisation and counter-urbanisation.
3. Describe two problems resulting from urban growth and explain proposed solutions.

1. See pages 132–3 2. See page 134 3. See pages 135–9

Chapter 11

Energy, resources and their management

LEARNING SUMMARY

After studying this section, you should be able to:

- ***define renewable and non-renewable resources***
- ***know the main sources of energy and their alternatives***
- ***understand how pollution and global warming are caused***
- ***define sustainable development***
- ***know the main principles of Agenda 21***

What are resources?

AQA A

When people use something from the Earth it becomes a **resource**. Until people started using petroleum and uranium they were not resources. Just as materials become resources so the process is reversible as materials are replaced by better ones. Coal has been replaced in many cases by oil.

Resources, energy and fuel

AQA A

Energy is used to run machines and to provide heat and light. Energy is stored in fuels, such as oil and coal, or in moving water and wind. Coal and oil are also raw material resources for the chemical industry. Today the main form of energy is electricity that can be derived from all of these fuels.

World consumption is already growing at a rate equivalent to the UK's entire energy market each year. This adds the equivalent of an entire USA market, currently the world's largest energy consumer, every ten years. Yet today two billion people are still without access to modern energy. As populations and cities grow, access to traditional energy forms becomes difficult, and local environments may be degraded.

The growing demand for resources and energy

AQA A AQA C

The demand for energy grows as population increases and economic development spreads to more countries. The demand for resources and energy is not evenly spread across the world. MEDCs have only 25% of the population of the world but they use up to 80% of the energy produced. Some MEDCs are trying to reduce their energy demand through more efficient motor vehicles and machines and better insulation. Most LEDCs get their energy from the efforts of people or animals or through imported fuels because they do not have significant **reserves of energy** of their own. Demand in LEDCs is set to rise as countries industrialise and more people own cars.

Renewable and non-renewable resources and energy

Renewable resources include hydro-electric, solar, wind, wave, geothermal and tidal power. They do not become exhausted.

Non-renewable resources include coal, oil and gas. They were formed millions of years ago and are known as fossil fuels. They can become exhausted. Currently about 95% of the world's energy supply comes from non-renewable sources.

Make a list of renewable and non-renewable resources. Be clear in your mind which resources can fit in either group.

Some resources can be thought of as both renewable and non-renewable. Wood can be used for fuel and the trees replanted. In practice in LEDCs the trees are cut for fuel and not replanted so they become non-renewable. Nuclear power is also non-renewable but current levels of reserves will last at least 1000 years.

Energy resources are vital to the development of a country.

Energy, resources and their management

AQA B AQA C

Coal

Coal is a sedimentary rock and was formed by the compression of trees and plants many millions of years ago. In the past it was the main fuel for industry, especially for iron and steel making, transport, particularly steam railways, and for the heating of homes. Today it is mainly used in power stations to produce electricity.

Large deposits of coal are found in Yorkshire, Derbyshire and Nottinghamshire, north-eastern France and Germany, the north-east USA, New South Wales in Australia, South Africa and India.

The main features of the **coal mining** industry are:

- it can be mined underground (shaft mining) or on the surface (open-cast)
- there are still 300 years of reserves for this relatively efficient fuel
- coal mines create visual, noise and air pollution
- mining can be dangerous and accidents do occur
- coal is bulky and costly to transport
- burning coal creates acid rain and adds to global warming.

Oil and natural gas

Fig 11.1 An oil field

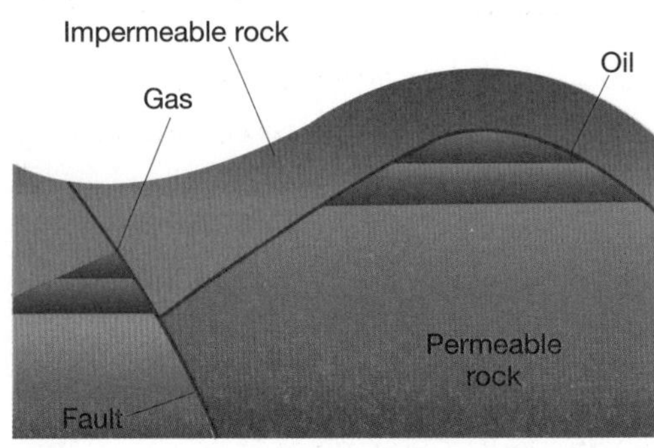

Oil and natural gas are formed from the remains of tiny sea creatures that died and were buried millions of years ago. Oil and natural gas are found in pockets beneath a layer of impermeable rock. The major oil fields are found in the Middle East, e.g. Saudi Arabia and Kuwait, Alaska and under the North Sea.

The main features of the **oil and gas industry** are:

- both fuels can be burned in power stations to produce electricity
- the fuels are extracted by drilling wells
- drilling can be dangerous, e.g. explosions and fire, in difficult environments such as Alaska and the North Sea

Note how OPEC controls the supply of oil and the effect this has on the economies of MEDCs.

- drilling in the North Sea takes place from massive oil rigs
- oil and natural gas are easier to transport than coal, using pipelines and massive sea-going tankers
- oil produces many valuable raw materials for other industries
- burning these fuels produces less greenhouse gases than coal
- known reserves of oil and natural gas will be used up by 2050
- serious environmental damage is caused if a pipeline breaks or a tanker is grounded, e.g. the *Exxon Valdez* in 1989
- developing an oil or natural gas resource creates wealth in the area
- oil has become a political weapon with the oil-producing and oil-exporting countries (OPEC) limiting supplies to maintain higher prices
- large reserves of oil are found in politically unstable areas, e.g. Iraq.

Nuclear energy

Nuclear energy is stored in uranium atoms and is released as heat to turn water into steam that drives steam turbines creating electricity. The energy in a nuclear power station is created when neutrons strike uranium 235 in a process called nuclear fission. Nuclear energy was seen as the source of energy for the future. **Accidents**, such as those at Five-Mile Island, Pennsylvania in 1979 and at Chernobyl in Russia in 1986, have caused many to question its use in the future. In the UK, nuclear stations were built on the coast at Hinckley in Devon, Sizewell in Suffolk, Hunterston near Glasgow, Dungeness in Kent, Dounreay in the north of Scotland and Wylfa on Anglesey. One problem for nuclear power stations is decommissioning at the end of their lives.

France has few fossil fuel resources and despite its hydro-electricity capacity has turned to nuclear power. Globally, there are about 500 nuclear reactors producing nearly 20% of the world's energy supply.

List the advantages and disadvantages of using nuclear power.

Views are divided on the future for nuclear power. The technology has the ability to create almost unlimited cheap power with relatively low environmental risks. Those against nuclear power point to the accidents, the dangers of military use, its inflexibility to meet peak demand and problems of nuclear waste disposal.

The main characteristics of the **nuclear energy industry** are:

- uranium reserves will last for many years
- refined uranium is relatively easy to transport
- power station sites do not need to be located near to the raw material source
- nuclear stations need large amounts of water and so tend to be located near to coasts or estuaries
- nuclear power stations produce no greenhouse gases
- it is easier for countries with low fossil fuel reserves to import small amounts of uranium to set up power stations
- there is widespread public concern about the safety of nuclear power both in terms of the transport of fuel as well as local radiation effects (unexplained clusters of cancer cases have occurred near some major nuclear stations)
- nuclear waste remains radioactive and dangerous for many hundreds of years.

Renewable sources of energy

AQA A AQA B AQA C

Using fossil fuels to generate energy does harm to the environment and these fuels will run out in the future. Nuclear energy is a politically unpopular source of power. For these reasons governments and non-governmental organisations (NGOs) are developing alternative sources of power. They aim to find and harness efficient, cleaner and more permanent sources of power. The main alternative sources of energy are hydro-electricity, the wind, the sun (solar power), the waves and the tides.

Hydro-electric power (HEP)

Fig 11.2 Cross-section through an HEP station

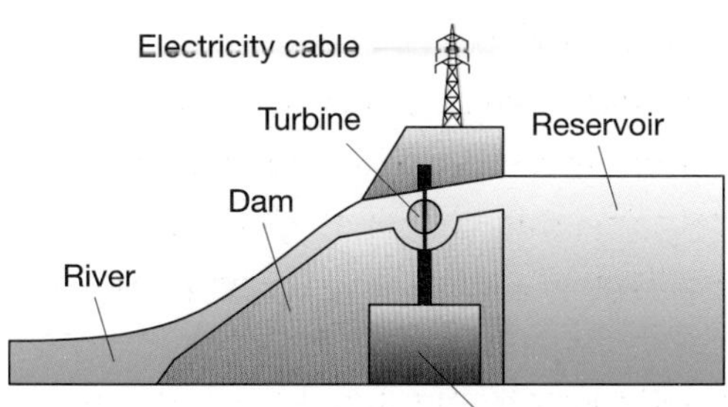

The power for **HEP** comes from moving water as it passes over a waterfall (Niagara Falls) or is released from behind a dam (Kariba Dam). The water drives turbines that generate electricity. Steep-sided valleys or gorges cut in strong impermeable rock in areas of reliable rainfall provide good sites. The huge scheme to dam the Yangtse in its gorge section has been criticised for the environmental damage it will cause with so many people losing their homes.

The main characteristics of **hydro-electric power** are:

- it does not produce any greenhouse gases
- once the dam is constructed, it provides relatively cheap electricity
- the dam controls the flow of water and can be used to prevent flooding downstream
- lakes behind the dam can be used for recreation and wildlife sanctuaries
- the dam is expensive to build and often difficult to access
- a large area of land may be flooded and people displaced (e.g. the Volta Dam in Ghana)
- the reservoir lake behind the dam can slowly silt up.

The consequences of increased resource use

Social consequences
• increases gap between the wealthy and poor • growing dangers to health • demand for energy increases to run 'essentials' such as cars and air conditioning
Economic consequences
• increase in prices as energy resources are depleted • increase in costs for exploration, development and production • need to find finance to develop renewable sources
Environmental consequences
• global warming • more pollution, danger of more accidents • fragile vegetation areas under threat, e.g. ice caps and forests
Political consequences
• need for international co-operation • changing power structure as energy rich nations emerge • public protests and lost votes when fuel prices rise

Pollution

Pollution topics to study:

- burning fossil fuels
- destruction of the rainforest
- freshwater and marine water pollution
- acid rain
- causes, consequences and strategies to alleviate pollution including international co-operation

Possible causes of and solutions to global warming

Global warming: the evidence

Rising temperatures and warmer weather since 1980 suggest that global warming is taking place. Some scientists predict a rise of 3°C in the next 100 years. Possible effects of global warming include:

- rising sea levels as the oceans expand at higher temperatures and the ice caps melt. Areas at risk would include the east coast of England, Bangladesh, the Nile delta and the Netherlands
- greater evaporation and increased rainfall
- increased plant growth where higher temperatures and water are available
- diseases such as malaria, yellow fever and dengue fever spread further north and south
- more changeable weather with stronger storms threatening coastal cities

Do not confuse global warming and greenhouse gases with the increased threat of skin cancer because of the increasing hole in the ozone layer from CFCs.

- low-lying areas such as Florida, the Netherlands and Lincolnshire under threat from rising tides and storms
- some cereal-growing areas and some poorer countries in Africa becoming drier
- loss of many birds and other animals including the polar bear as their habitat disappears.

No one can be certain that global warming is taking place. The higher temperatures could be another natural climatic variation, a fluctuation in solar output or the product of dust from volcanic eruptions.

The greenhouse effect

During the day the Earth is warmed by in-coming short-wave radiation from the sun (surface temperature of 5000°C). At night the Earth cools as it emits longer-wave radiation. Not all of the heat reaches space. Some is reflected back by clouds and some gases in the atmosphere which trap heat in the same way that a greenhouse would in a garden. Hence they are called '**greenhouse gases**'.

These greenhouse gases occur naturally and keep the temperature of the Earth 33°C higher. (Note during the Ice Age the temperature only fell by 4°C.) The amount of greenhouse gases in the atmosphere has been increasing as people burn fossil fuels, burn and clear rainforests, increase landfill sites, drive more miles, use refrigerators and aerosol sprays. This increase in greenhouse gases is trapping more of the outgoing long-wave radiation and reflecting it back to the Earth. More and more heat becomes trapped in the atmosphere of the Earth.

Sustainable development

 AQA A AQA B AQA C

 KEY POINT

Sustainable development meets the needs of the present generation and retains our ability to meet the needs of future generations.

Sustainable development can be promoted through conservation, resource substitution, recycling, use of appropriate technology, pollution control, using renewable energy sources such as wind and solar power, international action by groups of countries, and lower consumption of energy through better insulation, using fluorescent light tubes, using smaller cars, and cycling and using public transport instead of making car journeys.

Some planners in the UK have added to the definition of sustainable development. They include actions to ensure a settlement survives and is sustained. For example, they want to build affordable houses for young families in a village where the school is threatened with closure.

Agenda 21

In 1992 the **United Nations** produced Agenda 21 on sustainability with the slogan 'think globally and act locally'. The agenda has seven key points:

- interdependence – people, the economy and the environment depend upon each other locally and globally

- citizenship and stewardship – people have rights and responsibilities including participation and co-operation
- regard for the future – responsibility includes protecting future generations
- diversity – recognition and respect for the diversity of cultures, societies and environments on Earth
- equity – respect for people's quality of life with justice and fairness for all
- sustainable change – planning and monitoring change to ensure the future
- uncertainty – understanding the uncertainty of the future and taking appropriate precautions.

Five ways of using energy in a more sustainable way

Five ways of using energy in a more sustainable way
Conservation • protection of wildlife and scenery, e.g. through the management of National Parks • better insulation of homes • use of brownfield rather than greenfield sites
Resource substitution • using low cost energy schemes • using local building materials • using biodegradable materials
Recycling • collection by local authorities of paper, glass and plastics • replanting trees cut down for commercial purposes • reuse of building materials, e.g. grinding used bricks to make hard core
Pollution control • reducing pollution from cars and lorries • more efficient use of fuel in power stations • more checks and fines on those causing pollution
Using renewable energy • including wind, rivers, tides and the sun • grants to install equipment in houses • publicity on benefits and cost savings in the long term

Examples of **sustainable development** are:

- the use of brownfield land rather than green fields for new housing, factories and business parks
- the farming of forests so that more trees are planted than are being removed
- the imposition of quotas on sea fishing
- the provision of cycle tracks in urban areas
- improvements in electrically-driven public transport
- park-and-ride schemes (this is disputed by some who feel they encourage car use)
- organic farming using manure rather than chemical fertilisers
- the use of alternative energy sources instead of fossil fuels.

1. What is the difference between renewable and non-renewable resources?
2. Suggest reasons for the lack of popularity for the development of nuclear power.
3. Describe and explain what is meant by sustainable development.

1. See page 143 2. See page 144 3. See page 147

Chapter 12

Agriculture

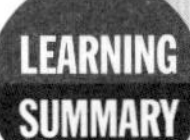

After studying this section, you should be able to:

- *understand farming as a system*
- *locate the different types of farming in the UK*
- *understand the factors affecting farming*
- *describe different types of farming*
- *recognise the ways in which farming is changing*

Farming, fishing, mining and forestry are called primary industries. They involve extracting resources from the Earth.

Farming as a system

AQA A AQA B

Each type of farming can be described as a system with inputs, processes and outputs (Fig 12.1).

Fig 12.1 Systems diagram of agricultural activity

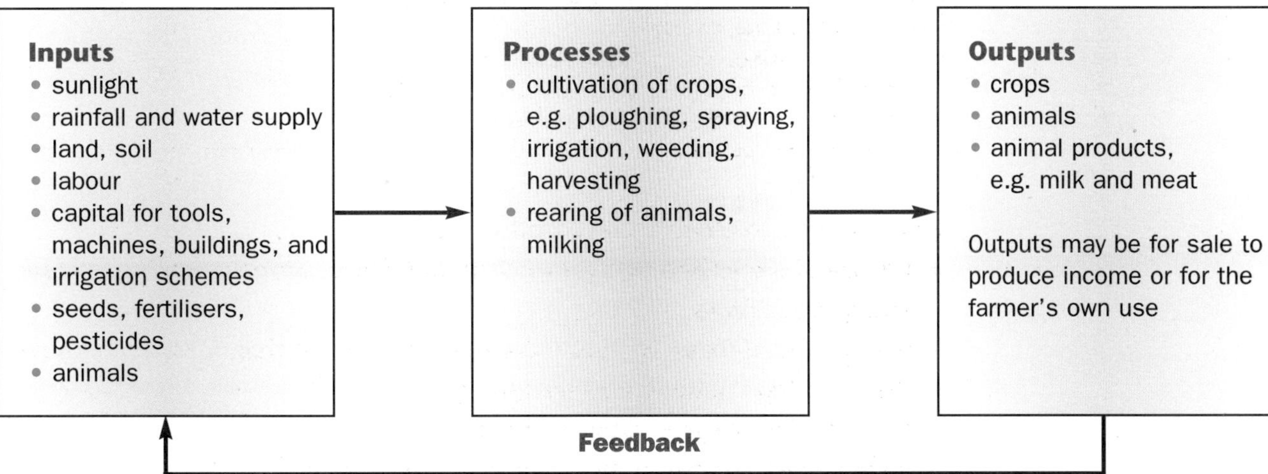

Feedback from the system

- Outputs produce profit for use on the farm
- Experience determines next year's decisions
- News from the government (NFU and MAFF) and media suggests changes
- Some output is retained as animal feed

Distribution of farming types in the UK: the four main types

AQA A AQA B

KEY POINT Each part of the UK tends to have a particular type of farming.

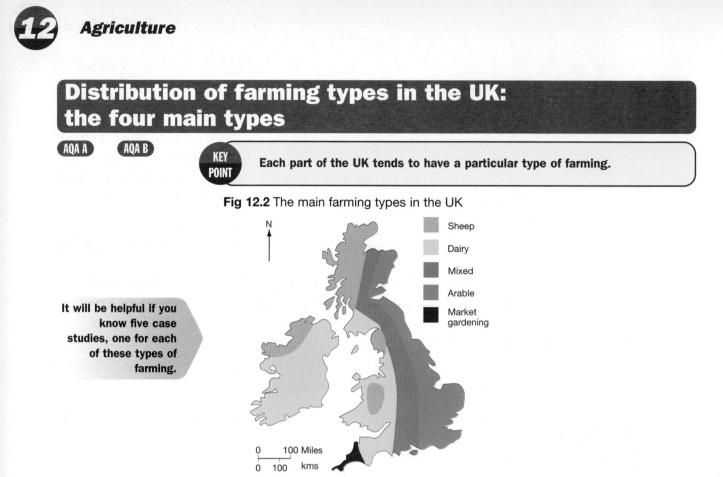

It will be helpful if you know five case studies, one for each of these types of farming.

Fig 12.2 The main farming types in the UK

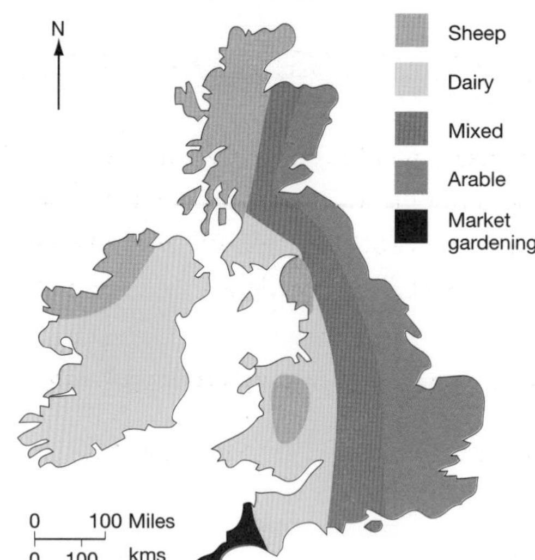

Dairy farming

This involves the rearing or purchase of cattle for the production of milk. It is mainly found in south-western England, the lowland areas of Wales and in Lancashire. Dairy farmers prefer fertile, well-drained soils that produce high quality grass. Dairy farms supply milk to nearby urban areas as well as to dairies for the production of milk products such as butter and cheese. The production of milk has been subsidised since 1945 but the quantity farmers are allowed to produce has been limited by **quotas** since the 1980s.

Arable farming

Arable farmers plough the land to produce cereal crops such as wheat and barley, vegetables, oil seed rape and linseed. Arable farms are mainly found in eastern England, including Norfolk and Lincolnshire, as well as the east of Scotland. The farmers use machinery and so prefer flatter land. An ideal climate has warm summers with rain during the summer growing season. Arable crops attract guaranteed prices through the **Common Agricultural Policy** (CAP) of the European Union.

Sheep farming

Sheep produce meat and wool. Sheep farms, which are mainly family run, are found in the upland marginal and peripheral areas of England, Wales and Scotland. Sheep farming can take place on steep slopes, with thin soils, high rainfall and low temperatures. The sheep feed on the poor grass that grows on the moorlands. Hill sheep farms receive subsidies from the EU and some farmers earn extra income from campers and caravan users. The lambs produced each February and March are taken to lowland farms to be fattened for sale as meat.

Market gardening

This is **intensive** farming, producing high-quality crops such as fruit, salad, vegetables and flowers. It can be found in Cornwall and Devon, the Isles of Scilly, and the Fens. Farms may be as small as one hectare. Market gardeners use climate-controlled glasshouses, artificial soils and irrigation to obtain a high quality output which is taken to markets in refrigerated lorries.

Factors affecting farming

AQA A AQA B

Farmers make decisions about what to grow, what animals to keep, the level and type of inputs and the methods they will use. Their decisions are based upon a range of social, economic and environmental factors. The farmers' attitudes and level of knowledge are also important.

Social and economic factors

The importance of each factor varies from farm to farm. Note how each influences farms near to you.

These are human factors and include labour, capital, technology, markets and government (political).

Labour

- In LEDCs, such as India and Java, farmers use abundant cheap labour instead of machines. In Japan and the UK, where labour is expensive, they use machines.
- People working on farms may be unskilled labourers or skilled and able to use machinery, e.g. tractors, harvesters and milking machines.

Capital (finance)

- Capital, the money the farmer has to invest in the farm, can be used to increase the amount of inputs into the farm, e.g. machinery, fences, seeds, fertiliser and renewing buildings.
- If a farmer can afford to invest capital, yields will rise and can create greater profits which can be used for more investment.

Technology

- Machines and irrigation are two types of technology that can increase yields.
- Greenhouses, with computer-controlled technology, provide ideal conditions for high quality crops. The computer controls the temperature, moisture level and amount of feed for the plants.
- Genetic engineering has allowed new plants to be bred that resist drought and disease and give higher yields.

Markets

- Farmers grow crops which are in demand and change to meet new demands, e.g. rubber plantation farmers in Malaysia have switched to oil palm as the demand for rubber has fallen.
- Markets vary throughout the year and farmers change their production to suit them.

Government

- Governments influence the crops farmers grow through regulations, subsidies and quotas.

Think of the farmer weighing up all these factors when deciding what to plant. Some richer farmers have plenty of choice.

- Governments offer advice, training and finance to farmers and, in new farming areas, may build the infrastructure of roads and drainage, e.g. Amazonia.
- In some countries, e.g. Kenya and Malaysia, the government is trying to help nomadic farmers to settle in one place.
- Some governments plan and fund land reclamation and improvement schemes.

Environmental factors

These are **physical factors** and include climate, relief and soil.

Climate

- Temperature (minimum 6°C for crops to grow) and rainfall (at least 250mm to 500mm) influence the types of crops that can be grown, e.g. hot, wet tropical areas favour rice, while cooler, drier areas favour wheat.
- The length of the **growing season** also influences the crops grown, e.g. wheat needs 90 days. Some rice-growing areas have two or three crops per year.

Relief

- Lowlands, such as flood plains, are good for crops.
- Steep slopes hinder machinery and have thinner soils; lower, more gentle slopes are less prone to soil erosion.
- Tea and coffee crops prefer the well-drained soil on hill slopes.
- Temperature decreases by 6.5°C for every 1000 metres gained in height.
- South-facing slopes receive more sunlight.

Soil

- **Fertility** is important; poor soil means lower outputs or larger inputs of fertilisers.
- Floodplains are good for rice because of the alluvial soils.
- Good drainage reduces the dangers of waterlogging.

1. Name the five main types of farming in the UK and for each identify two main characteristics.
2. Describe the difference between social and economic factors and environmental factors influencing farmers' decisions.
3. Describe how governments can affect farming in LEDCs and MEDCs.

1. See pages 150–1 2. See pages 151–2 3. See page 151–2

Different types of agricultural activity

AQA B

There are many different types of agricultural activity including:

- commercial farming, e.g. dairy farming in Worcestershire, cereal crops in France (EU) and plantation agriculture, e.g. west coast of Malaysia for oil palm and tea
- subsistence farming, e.g. shifting cultivation in the Amazon Basin of Brazil and Sabah and Sarawak in Malaysia
- intensive farming, e.g. market gardening in the Vale of Evesham, flowers and bulb cultivation in the Netherlands and wet rice cultivation on the Ganges Plain of India

Use an atlas to locate the many places named in this section.

- extensive cultivation, e.g. cereal growing in Norfolk, sheep farming in Australia.

Subsistence farmers grow crops and rear animals mainly for their own use.

Case Study I: Shifting cultivation in an LEDC

Amerindians in Amazonia

This is very helpful for AQA B.

Inputs	Low inputs Forest clearance and burning Manual labour and simple tools Some groups have chain saws and axes
Outputs	Crops for the family to use for food and clothing Charcoal Rice, maize, tapioca, sweet potatoes, bananas and vegetables Low yields as soil rapidly loses its fertility
Physical factors	Cultivation of remote areas of equatorial forest High rainfall, high temperatures, high humidity Dense vegetation and flooding
Human factors	Many crops grown, up to 30, in small clearing Governments are trying to persuade shifting cultivators to settle on farms Low educational levels, people vulnerable to disease
Processes	Crops are grown by members of the family Low technology, mainly axes and knives to burn and clear dense forest and thick undergrowth Slash and burn agriculture causes areas of forest to be cleared
Problems	Land is abandoned quickly and suffers soil erosion Decreasing areas of forest as ranching and logging expands Population increase is reducing areas available for shifting cultivators

Fig 12.3 Distribution of shifting cultivation

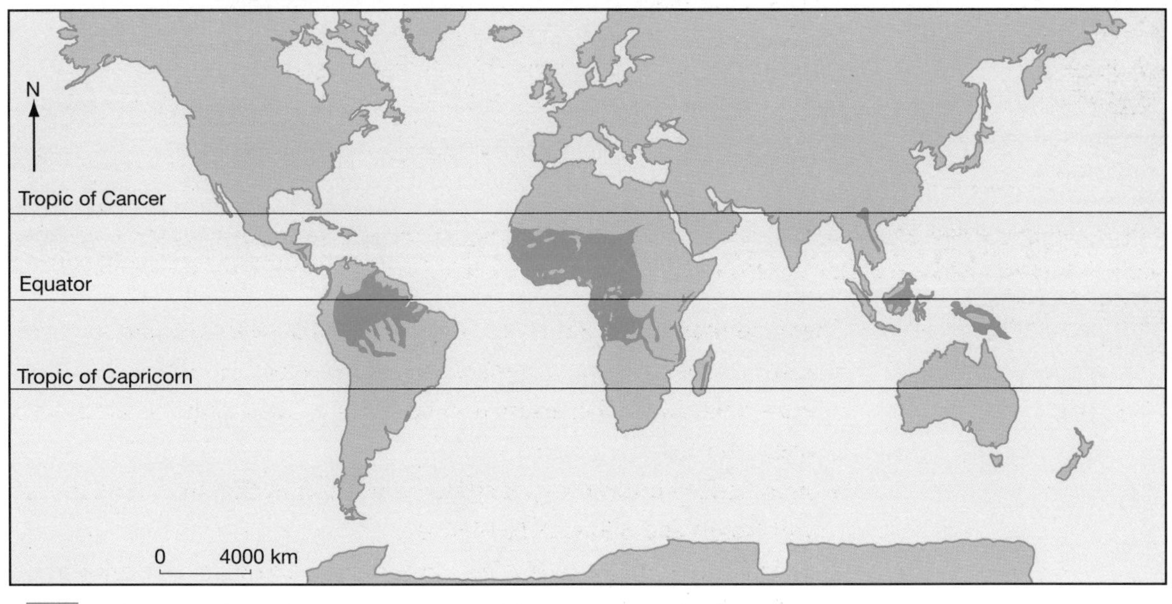

Case Study II: Intensive farming in an LEDC

Rice growing in the Ganges Plain of India

This is very helpful for AQA B.

Inputs	High inputs of fertiliser or more fertile silt as river floods Large amounts of labour
Outputs	One third of the population of the world depend on rice for food Some vegetables are grown Many small farms only grow enough rice to feed the family
Physical factors	Rice needs between 1000mm and 2500mm of rainfall each year Average temperatures of 20°C Fertile alluvial soil (silt) found in the Ganges Basin Soils are replenished each year by silt added during monsoon floods
Human factors	Low levels of technology, manual labour and water buffalo Small farms, fragmented by inheritance laws that demand that land is divided between sons Ganges Plain is densely populated Rice has a high nutritional value Much labour is needed to build 'bunds' (banks) around padi fields
Processes	Irrigation often allows more than one crop per year Planting and transplanting usually done by hand Intensive subsistence farming of rice
Problems	Many farms are too small to provide any surplus that can be sold in the market Farmers are too poor to invest in ways to improve their farms High yielding varieties of seed for rice need fertilisers that are too expensive for poor farmers to buy

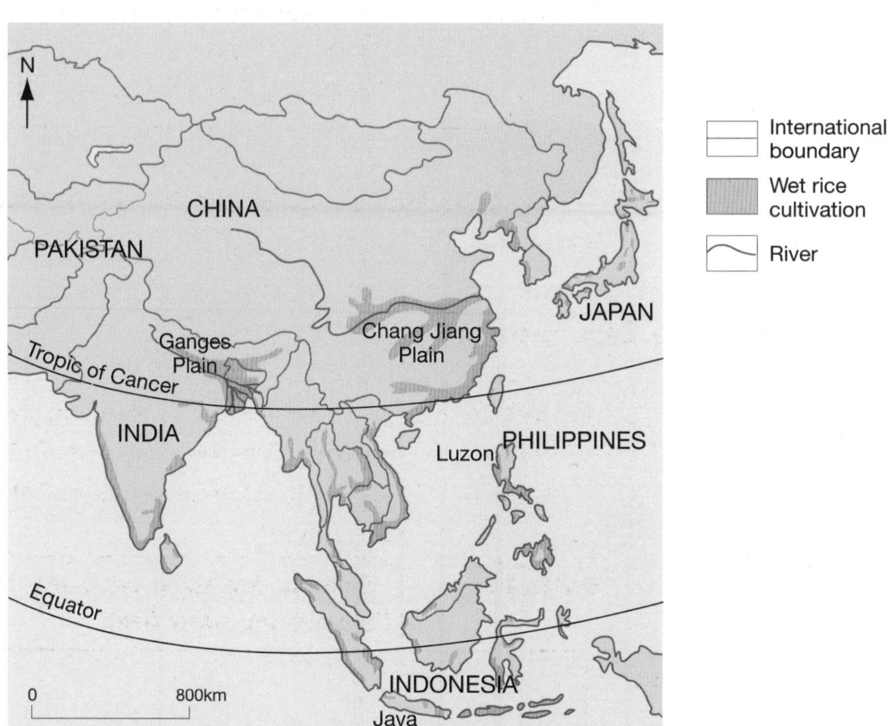

Fig 12.4 Wet rice cultivation in Asia

Commercial farmers grow crops and rear animals for sale to make a profit. Intensive farming uses high inputs, such as labour and capital, per unit area.

Case Study III: Intensive commercial farming in the EU

Farming in Denmark

Inputs	High inputs per hectare of fertiliser Seeds and animal breeding
Outputs	Wheat, oats, barley and sugar beet grown for animal food or brewing Dairy produce – milk, cheese Meat and bacon
Physical factors	Poor soils have been improved by the addition of manure and fertiliser Large fields bordered by wire fences, no hedges, to increase area for cultivation Climate encourages cereal growing rather than grass
Human factors	Farms tend to be built around a courtyard, farmhouse on one side, animal and machinery sheds on the other sides Smaller farms disappearing, more mechanisation, fewer farm jobs
Processes	Animals kept indoors from September to April (seven to eight months), fed on crops grown on the farm Farmers work in groups, called co-operatives, to share storage, machinery, transport and marketing costs Co-operatives bulk buy seeds and fertilisers, run a dairy and make cheap loans to farmers Crops are rotated, usually on an eight-year cycle
Problems	Climate does not allow good quality grass to be grown Shortage of labour for twice-daily milking Price changes as too much milk is produced in the EU

Case Study IV: Extensive commercial farming in the UK

Cereal growing in East Anglia

Inputs	Low rain – falls mainly during the summer growing season Warm, sunny summers that aid the ripening of the grain Winter frosts break up the soil after it has been ploughed in the autumn
Outputs	Wheat, barley, sugar beet, peas and beans Potatoes and sugar beet

Physical factors	The land is gentle, undulating, allowing easy use of machinery Soils are well drained, e.g. chalk or alluvium based
Human factors	Roads and rail systems are well developed Large markets available in nearby large cities, e.g. London Large farms tend to be owned by wealthy farmers or large companies
Processes	The year begins in the autumn with ploughing, harrowing and sowing. During the spring, farmers weed, apply fertiliser and pesticides. Harvest takes place in late summer. Farmers also have their field boundaries, buildings and machinery to maintain. Farmers rotate their crops to maintain the quality of their soil
Problems	Soil eroision as large fields with fewer hedgerows provide less protection Dangers of water pollution if fertilisers applied incorrectly Political factors and changes in EU policy make planning difficult Cheaper imports push down prices Making plans for set-aside land Need to diversify in order to maintain income

Case Study V: Plantation agriculture

Inputs	Large amounts of capital for road building, planting and processing factories, schools, hospitals and houses Large numbers of people employed in labouring, administration and technical services Capital is also needed to bridge the gap between planting of trees and the first harvest Fertilisers and pesticides are used to improve output
Outputs	Crops grown for export or use by local industry Massive output of one crop Tea, coffee, spices, oil palm and rubber The aim is to sell the crop for cash
Physical factors	Usually takes place on large farms or estates, e.g. 40 to 1000 hectares Hot and wet climate encourages plantation crops Rubber trees grow well on the slopes of the foothills in Malaysia
Human factors	Estates were often developed in the eighteenth and nineteenth century colonial period. European countries such as the Netherlands and UK set up plantations to grow tea, sugar, spices, coffee and rubber Large employers of local people who have skills to check crop quality Plantations raise the standard of living of local people Some plantations have been taken over by the Malaysian government

Processes	Trees are first planted in nurseries and then transplanted to production areas Crops exported to Europe to meet growing industrial demand High levels of technology used, such as machines for planting and harvesting
Problems	Owners and managers tend not to be local people and some money leaves the country Many plantations are owned by large multinational companies Fluctuations in world prices and demand for the products Soil erosion is present on some plantations Need for land for local people means that some plantations are being broken up to provide smaller farms

Agricultural change

AQA A AQA B

KEY POINT **The Green Revolution involves the use of high-yielding varieties of seed, increased irrigation and the use of fertilisers to improve food supply.**

Contemporary solutions to problems of farming regions in LEDCs

Meeting the rising demand for food: the Green Revolution

During the 1960s, as world population grew, there was increasing concern in many countries about providing an adequate **food supply**. The main plan to increase food supply was called the Green Revolution.

The Green Revolution can be included in questions on agriculture or food supply. It is very popular with examiners.

Countries such as India and Indonesia made great efforts to increase the supply of rice and wheat using irrigation, new seeds and fertilisers. The main aspects of the **Green Revolution** are:

- use of high-yielding varieties (HYVs) of seed produced by scientists in the Philippines. The first successful HYV was IR-8 the 'miracle rice' that doubled yields
- larger amounts of fertiliser are needed by the HYVs
- HYVs need adequate and carefully-controlled water supplies
- further development of HYVs has reduced the growing period from 180 to 100 days
- scientists are continuing to develop new varieties of seed that need less irrigation, are resistant to disease and are good to eat
- governments are offering loans, advice, storage and transport facilities to poorer farmers.

The Green Revolution has brought benefits and problems.

Benefits

- Increased rice production; the total produced doubled, especially in China.
- Improved standard of living as farmers sold their surplus, e.g. the Punjab in India where farmers were willing to accept the changes.

Problems

- Irrigation is essential for the best results from HYVs.
- HYVs are not successful on alluvial plains where flooding occurs.
- HYVs are more costly to grow, needing more fertiliser and irrigation.
- Farmers who can afford to grow HYVs get richer and this increases the gap between them and the poorer farmers.
- Fertilisers for HYVs are creating pollution in rivers and lakes.
- HYVs have been vulnerable to outbreaks of disease.
- Mechanisation has increased unemployment.
- Intense use of land increases dangers of soil erosion.
- Overuse of water from irrigation creates waterlogging and salinisation.
- Sophisticated technology may be inappropriate for local people, e.g. expensive electrical water pumps cannot be maintained by local craftspeople but simple ones can.

Use a table to list advantages and disadvantages.

Contemporary solutions to problems of farming regions in MEDCs

This is very helpful for AQA B.

The European Union (EU) and its Common Agricultural Policy (CAP)

The CAP is unpopular and seen as protecting poorly-organised farmers at the expense of the public. It protects European farmers against cheaper food from LEDCs.

The **CAP** has been in action, for member states including the UK, since 1962. It aims are:

- to protect the income of farmers
- to ensure reasonable and steady prices for consumers
- to ensure food supplies through increased production
- to protect the quality of life in rural areas.

These policies have been put into action through grants, subsidies and guaranteed prices. Stable prices allow farmers to plan. The policy has been criticised for protecting weaker farmers, creating mountains of food and lakes of wine, making food expensive when compared with the USA and for environmental damage (hedgerow removal, soil erosion, excessive use of fertilisers, pesticides and herbicides) as farmers attempt to increase production. Recently the EU has been paying farmers to take part of their land out of production (set-aside).

Farm diversification

The National Farmers Union has produced a list of nearly 30 ways in which farmers can use their land to increase their income.

Farmers in the UK have found it increasingly difficult to make their farms economic. Food surpluses, milk quotas, reduced subsidies and growing public awareness of environmental issues have combined to reduce farm income. Farmers have been encouraged by government to move to other means of earning income (diversification).

Farmers have turned to:

- pick-your-own (PYO) for apples, strawberries, gooseberries and other fruit
- garden centres and farm shops for the sale of farm produce and garden plants
- barns and outbuildings converted to holiday cottages
- small industrial estates near to the farm buildings
- golf courses and nature reserves open to the public on farm land.

1. Describe the main inputs to a farm.
2. What are the five main types of agriculture found in the UK?
3. List the human and physical factors influencing a farmer's decisions.

1. See page 149 2. See page 150 3. See pages 151–2

Chapter 13

Industry

LEARNING SUMMARY

After studying this section you should be able to:

- *define primary, secondary and tertiary industry*
- *understand industry as a system*
- *know the main factors of industrial location*
- *describe industry in LEDCs and MEDCs*
- *understand the rise of the Newly Industrialised Countries (NICs)*
- *understand the changing structure of industry*
- *recognise change in the location of shops*

Different types of industry

Industry can be divided into three main types: primary, secondary and tertiary industry. Recently, tertiary industry has been sub-divided to give a fourth type: quaternary industry. Sometimes secondary industry is called manufacturing industry and tertiary is called services.

An occupation is the job that someone does; employment is the industry in which they work. A person can have the occupation of accountant and be employed in a car assembly plant.

A triangular graph is a useful way of showing the balance of primary, secondary and tertiary industry in a country.

Primary industry

Primary industry involves extracting resources from the sea or land and includes farming, fishing, forestry, coal mining, oil drilling and hunting. It is located where the raw material is available. Primary industries are particularly dependent on physical factors such as climate and geology as well as government intervention, e.g. the Common Agricultural Policy of the EU.

Secondary industry

Secondary or manufacturing industries make things for people by processing raw materials or assembling components. The raw materials may be obtained from primary industry or be the products of other secondary industries. A tin of fruit or a motor car are manufactured by a secondary industry using the products of primary and manufacturing industries.

Tertiary industry

Tertiary industries provide a service and include doctors, teachers, shop assistants, entertainers and lawyers. Health, administration, retailing and transport are called service industries.

Quaternary industry

Quaternary industry is concerned with research and development as well as information and communications technology which help companies to function.

Industry as a system

AQA B

Manufacturing industry can be seen as a system with inputs, processes and outputs (Fig 13.1).

Fig 13.1 Systems diagram of manufacturing industry

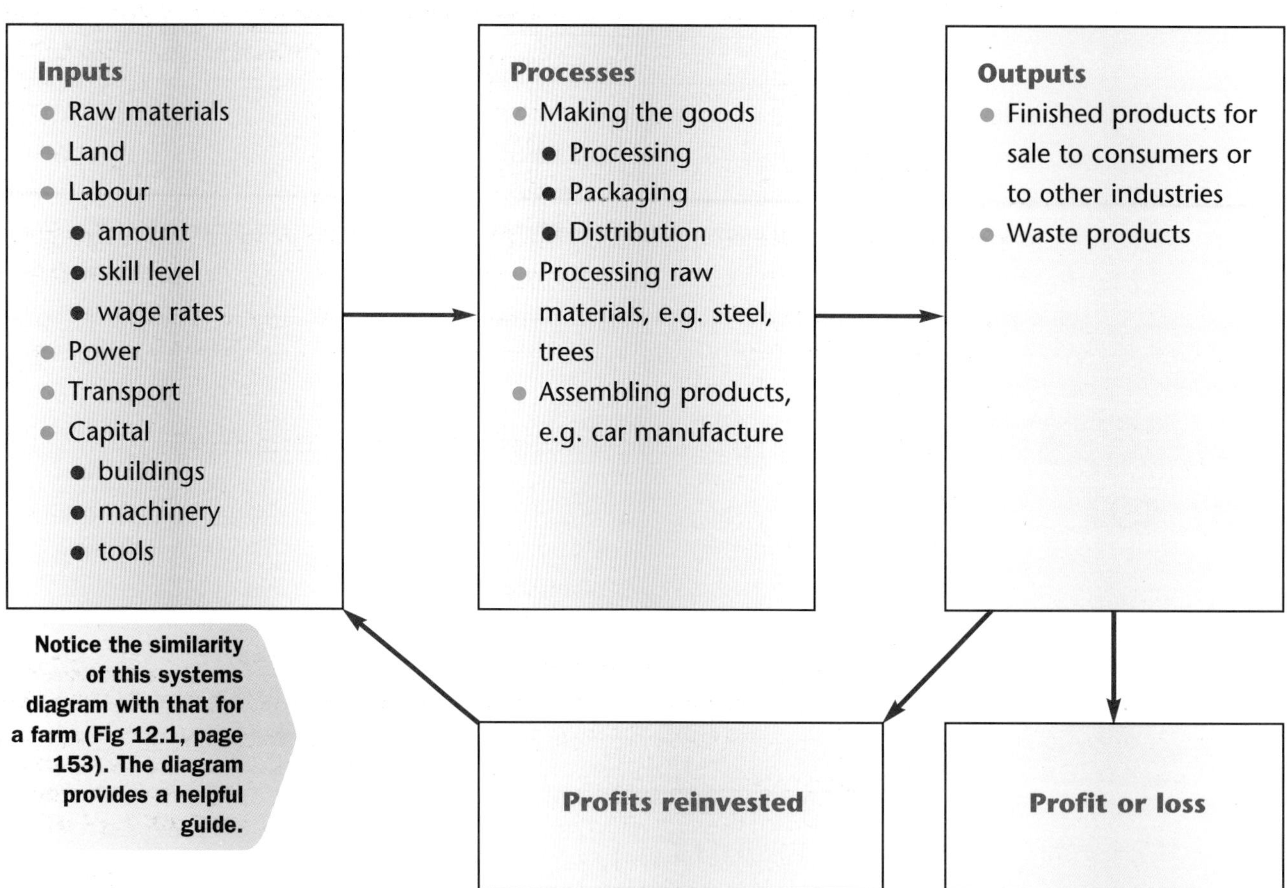

Notice the similarity of this systems diagram with that for a farm (Fig 12.1, page 153). The diagram provides a helpful guide.

Manufacturing systems are called 'open systems' because only part of the income from sales is used for inputs and reinvestment. Some is taken by the owners or shareholders as profit.

PROGRESS CHECK

1. Divide the following occupations into primary, secondary and tertiary industry:

Coal miner	Dairy farmer	Taxi driver	Doctor
Comedian	Teacher	Bricklayer	Garden designer
Tax collector	Cabinetmaker	Hairdresser	Laboratory technician
Waiter	Journalist	Shopkeeper	Factory worker

2. For an industry or factory you have studied, draw a systems diagram.
3. Make a list of at least three occupations found in your local area for each of the four types of industry. Present your list as a table.

1.

P	P	T	T
T	T	S	T
T	S	T	S
T	T	T	S

Factors affecting the location of industry

KEY POINT **Secondary industry or manufacturing industry involves the processing of raw materials or other manufactured products to produce goods.**

Fig 13.2 Spider diagram of the factors affecting the location of industry

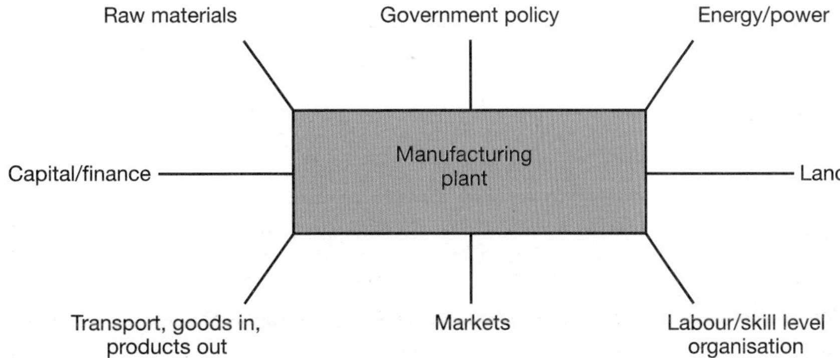

Industrial location factors for manufacturing industry

Fig 13.3 Pull factors on the location of a factory

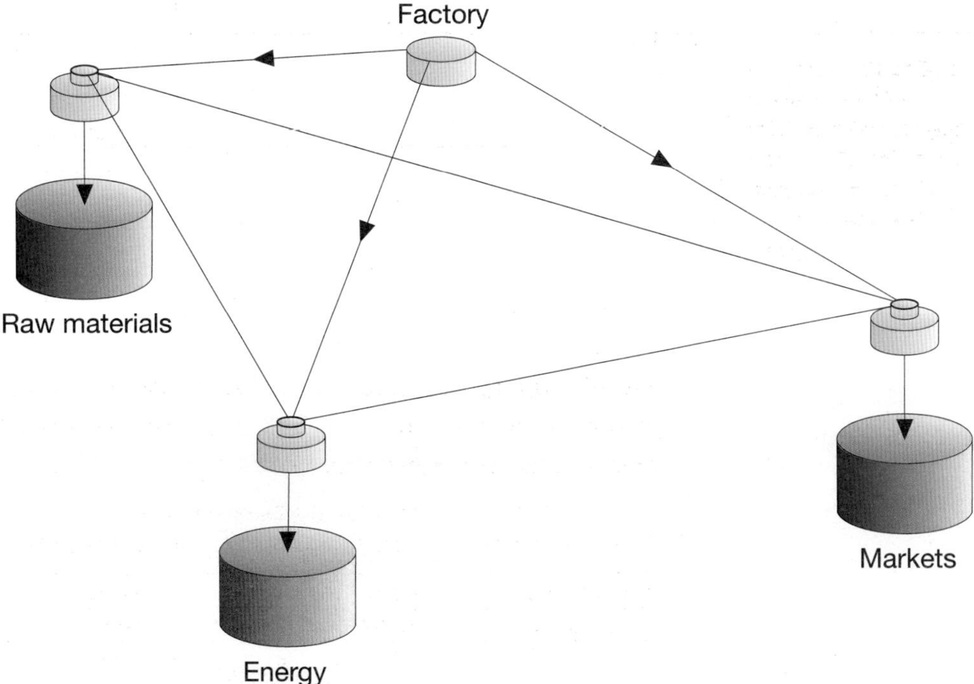

Raw materials

These are essential for the production of goods. In the past they have been very important in the location of industry. Some raw materials are very bulky or heavy and were expensive to transport. As a result, factories tended to be located near to the raw material. This was true of the iron and steel industry where the raw materials, iron ore and coal, were bulky and used in large quantities. Costs were reduced if the iron and steel plant located near to iron ore and coal deposits. The steel produced, though heavy, was less bulky and

Be able to list ten factors that influence the location of industry.

easier to transport than the raw materials. The iron and steel industry in the Ruhr, in Germany, was a good example of this factor. Another example would be the manufacture of cement in the Peak District of Derbyshire. Several tons of limestone are used to manufacture one ton of cement. As a result the manufacturing plant is next to the limestone quarry. These industries can be described as 'weight losing'.

Some products **gain weight or bulk** during manufacture. Baking, brewing and soft drinks are typical of this type. A bag of flour is compact and produces a much greater volume of bread. Beer and soft drinks contain large quantities of water. These 'weight-gaining' industries used to locate near to their markets to reduce transport costs.

Energy or power

Industries need **power** to operate their machinery. In the past, industries relied upon steam power and tended to be located near to coalfields. Today they use **electricity** that is generated in power stations and transmitted over long distances using the **National Electricity Grid**. This has released most industry from the need to be close to a source of power.

The manufacture of aluminium takes very large amounts of power. As a result some aluminium manufacturing plants have located close to sources of hydro-electricity power to reduce their energy costs.

Land

Larger industrial plants, such as the steel industry, prefer flat land. This makes it easier and less costly to build and operate the plant. It is also helpful if the site provides space for expansion in the future.

Capital or finance

For an industry in your area list the factors influencing its location.

To set up, run and improve an industry takes large amounts of money. Even a small factory will need some capital to get started. All types of industry have to purchase raw materials, machines, premises and transport equipment. Banks, finance houses and governments offer loans to people (**entrepreneurs**) developing a business. These financial investors demand to see a plan of how things will develop and expect some return for the money they have invested. In some cases the source of capital may come from abroad.

Labour

The **availability**, **skill level**, **cost** and **working practices** of the workforce are all important factors in the location of industry. For example, the textile industry needs large numbers of workers who need not be highly skilled. As a result textile factories tend to locate near to or in large urban areas. In the nineteenth and first half of the twentieth centuries, Lancashire and Yorkshire had many textile manufacturing factories. Most of these have closed, as cheaper labour has now become available in South-East Asia.

Other industries need pools of educated and skilled labour that can be easily trained for new technologies. A good example is the computer hardware and software industries that originated in Silicon Valley in California but have now moved to the newly industrialised countries (NICs) like South Korea and Taiwan.

Productivity and worker attitudes are also important to employers looking to establish new factories. An area that is known for its stoppages, strikes and poor workmanship will not attract employers.

Factories need to be managed. When a site for a factory is chosen, the views of the owners and managers are important. They will have a view about the environment in which they wish to live and work. They may consider the quality of education available as well as leisure facilities such as golf courses.

Transport

Transport is needed to bring raw materials or the products of other industries to the factory and to deliver finished products to the market or other factories. A good transport network will help move products quickly. A site that is well connected to a **motorway network** with rail and airport connections will be more attractive. Sites in more remote or peripheral regions will have higher transport costs.

Markets

How have changes in one of these factors changed the location of industry, e.g. improvements in transport?

The goods produced are sold to meet the demand of the market. Sometimes people in the market will order goods from a factory. At other times, salesmen have to go into the market to persuade people to buy the product.

Industries that produce perishable goods will tend to locate near to the market. However, developments such as refrigeration have made this less necessary.

The people living in an area or country provide a domestic market. If they are affluent, their spending will encourage the growth of industry. In poorer LEDCs there is only a small domestic market and this hinders economic growth.

Where a factory sells nearly all its product to a small group of factories it is usual for the factories to group together. This is said to create an agglomeration and is an example of **industrial linkage**.

Government policy

There are a number of ways in which governments may influence the location of industry including:

- setting aside plots of land for industrial estates
- banning industries that create pollution from siting near to residential areas
- offering loans, subsidies and tax exemptions to companies willing to open factories in regions in need of development
- providing stable government, without corruption, so that overseas investors are attracted to the country.

Industrial inertia

A study of an established factory or industry in an urban area might show little obvious reason for its location. This may be because the original factors that gave it an advantage on the site no longer apply. However, despite the disappearance of these advantages, the factory continues to be successful. This may be because of its reputation. When this happens it is said to be a case of industrial inertia. If the factory were to be located today, another site with up to date advantages would almost certainly be chosen.

The special case of the footloose industries

Many of these factors remain important in understanding the location of industry. However a growing number of industries are free to locate in many locations. These are called the footloose industries. They include many of the growing industries.

1. List the eight factors to consider when studying the location of industry.
2. What do you understand by the term 'industrial inertia'?
3. How can governments influence the location of industry?

1. See pages 165–168 2. See page 168 3. See page 167

Economic activity in MEDCs and LEDCs

AQA A AQA B

There are marked differences between the economic development of MEDCs and LEDCs. However for many LEDCs the gap is shrinking as transport and communications make links across the world easier.

There are also different levels of economic development within countries. Southern Italy and South Wales contrast with their economically prosperous areas in northern Italy and south-east England respectively.

Japan, Brazil and Bangladesh are countries at different levels of economic development.

Industry in LEDCs and MEDCs

AQA B AQA C

Case Study I: Industry in LEDCs

The development of industry in Malaysia

Inputs	In Malaysia the government has overcome its problems and attracted transnational companies (TNCs), sometimes called multinational companies Government has attracted TNCs with lower taxes, freedom from tariffs and generous quotas TNCs have headquarters, research, development and marketing section in their parent MEDCs

Know why TNCs locate in LEDCs and what advantages and disadvantages they bring.

	TNCs bring with them foreign exchange (FOREX) to be used as capital to buy machinery for new factories TNCs train local people in new industrial skills as well as building roads, schools and hospitals
Outputs	Originally the manufacture of agricultural products and then the growth of heavy industry More recently governments have switched to high-tech industries
Physical factors	Land available for the development of industry Location on the rim of the Pacific Ocean near to many other populated countries
Human factors	Lack of capital to invest in industry Investors from MEDCs reluctant to invest in LEDCs because of uncertain conditions Shortages of skilled labour and management expertise Unstable and corrupt governments TNCs are attracted to Malaysia by low labour costs Employment of local people creates growing domestic market Governments ban strikes so that production will not be interrupted The government has encouraged migrant workers to come from other countries to solve problems of labour shortages
Processes	New factories on industrial estates purpose built for modern industry with flexible layouts Ports and airports have expanded to link with the rest of the world Heavy industry was located close to sources of raw material or near to ports Unreliable sources of power
Problems, solutions and environmental impact	Poor transport systems that are vulnerable to the weather Limited local markets because of low incomes locally Unfair trading by MEDCs including quotas on imports With the help of TNCs, Malaysia has developed rapidly into an industrialised country and is called an NIC (Newly Industrialised Country). South Korea and Taiwan are also NICs and are said to have 'Tiger' economies TNCs bring disadvantages including: • they pressure governments to make decisions in favour of them rather than the country • they transfer profit back to their home country • many of the jobs they create are of low skill • wages can be low and working conditions not good • they may cause environmental damage • they have their own plans and may leave the LEDC for an even cheaper location Recently, in Malaysia the government has privatised much of its industry to allow for more international support and growth

It will be helpful to know a case study of one NIC.

Case study II: Industry in MEDCs

Traditional heavy industries in the Ruhr Valley of Germany

This is very helpful for AQA B.

Inputs	In the past the Ruhr had local raw materials including coal, iron ore and limestone Imports of raw materials are brought into the region via Rotterdam and the Rhine The steel industry has only survived through subsidies from the government Newer industries are moving into the region to replace the older traditional ones Companies like Hitachi and Mitsubishi have overtaken the older German names such as Krupp and Hoesch
Outputs	The Ruhr developed a heavy manufacturing industry producing steel, chemicals and textiles Some its importance as a heavy industrial area has been lost in recent years The output of equipment for the electronics, telecommunications and the media industries has increased
Physical factors	Sources of coal, iron ore, limestone and water supply from local rivers Rivers also provided a source of transport for the heavy goods that were produced Many of the original raw materials are now exhausted The Ruhr is at the centre of a huge European market and its location has attracted many new companies
Human factors	The region developed particularly in the late 1930s and 1940s to supply arms for Germany Labour costs are high in order to support high living costs within Germany. Steel is much cheaper from Taiwan and South Korea despite the long distance that it has to be transported (this is described as the globalisation of industry) Workers have had to learn new skills as the nature of the industry has changed
Processes	The production of iron and steel using blast furnaces has been largely replaced by lighter industries Many of the newer companies make component parts for other industries or assemble the components into the finished product
Problems, solutions and environmental impact	Since 1960, many of the large steel manufacturers have decreased in size and many jobs have been lost in cities such as Dortmund and Bochum Many of the factories are old and out of date The new factories take account of modern working conditions and the need to protect the environment In the remaining heavy industrial plants many of the waste products, such as heat, are recycled Much of the derelict land has been restored for the use of housing, industry and parks

Fig 13.4 Main processes in an integrated iron and steelworks

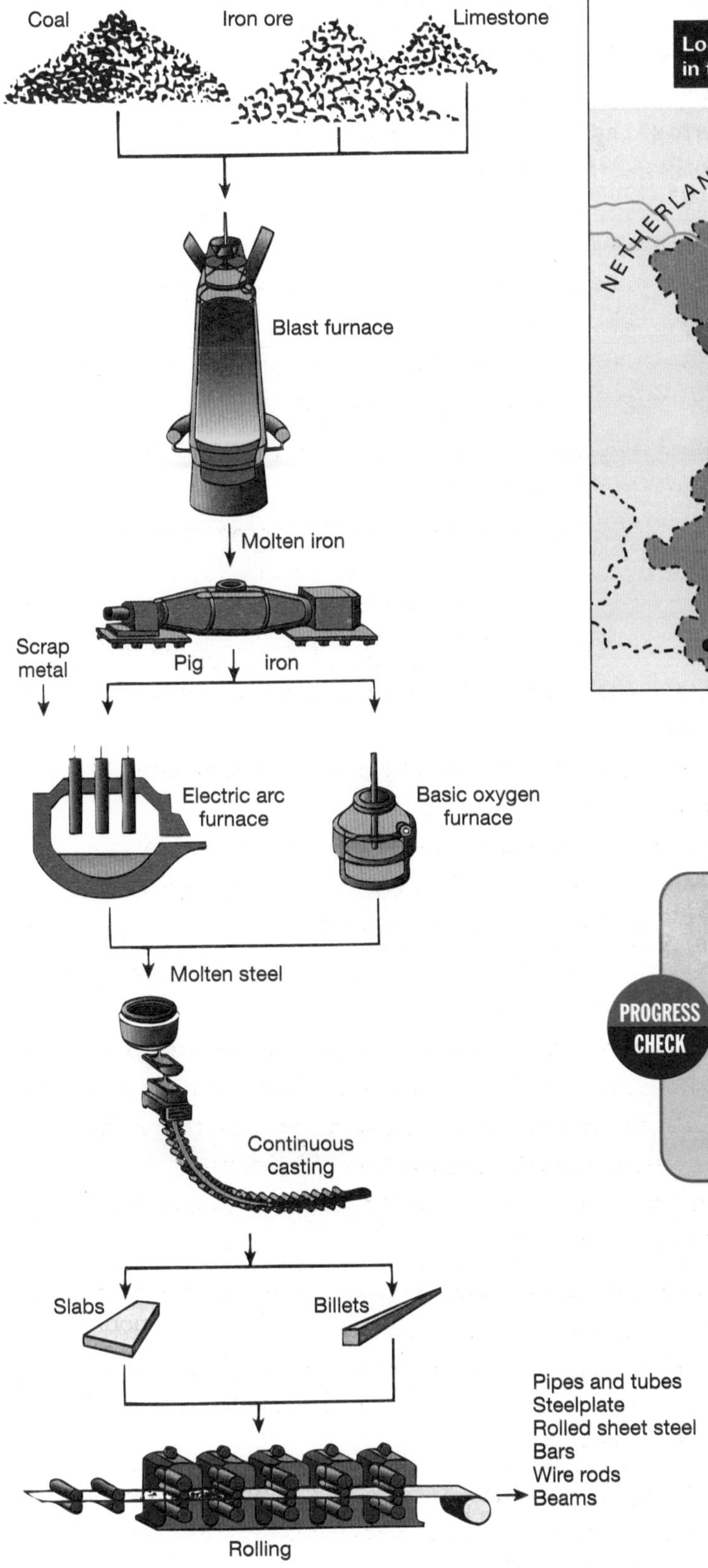

Coal

Iron ore

Limestone

Blast furnace

Molten iron

Scrap metal

Pig iron

Electric arc furnace

Basic oxygen furnace

Molten steel

Continuous casting

Slabs

Billets

Rolling

Pipes and tubes
Steelplate
Rolled sheet steel
Bars
Wire rods
Beams

Fig 13.5 Map of the Ruhr

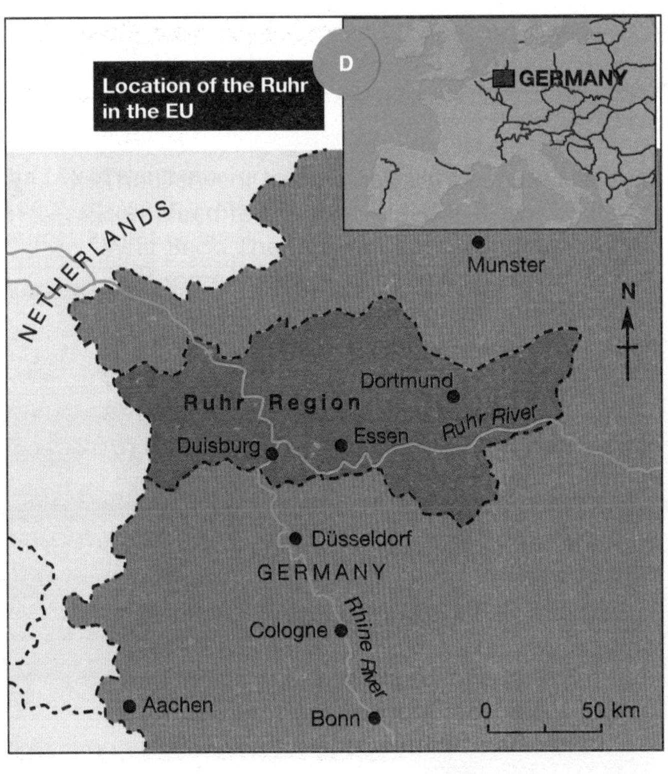

Main industrial area

City

PROGRESS CHECK

1. What is meant by the letters TNC, NIC and FOREX?
2. Why do LEDCs wish to industrialise?
3. What different problems face industries in LEDCs and MEDCs?

1. Transnational Company, Newly Industrialised Country, Foreign Exchange 2. See page 164
3. See pages 164–6

Case Study III: New industrial areas in the UK

High-tech industry in the M4 corridor

This is very helpful for AQA B.

Inputs	Growth of modern factories on new industrial estates stretching from the western edge of London, near Heathrow airport, to Bath and Bristol and including Reading, Newbury, Swindon and Bath Universities in London, Oxford, Reading, Bath and Bristol provide highly educated graduates The high-tech industries tend to be the new growth industries (sometimes called the 'sunrise' industries) Famous names such as Dell, Hewlett Packard, Sony, Bosch and Oracle are located in the area Raw materials are small and can be transported easily Investors from many countries are ready to invest in the area
Outputs	Research and development of new electronic products Computers, computer software, telecommunications and media equipment
Physical factors	Rapid road communications along the M4 motorway Large areas of flat land for new factories usually on the edge of cities Large areas of the industrial estates are covered in grass and trees, making them attractive places to work Fast railway service along the corridor The environment is attractive with the Cotswolds, Mendip Hills and water parks nearby
Human factors	These companies tend to employ relatively small numbers of highly skilled people The workforce is intelligent and inventive; they tend to move around from company to company Wages tend to be high and this creates high demand for service industries
Processes	Industries are said to be 'footloose' because they are not controlled by the traditional industrial location factors Some firms are based on older manufacturing companies such as Rolls Royce and British Aerospace
Problems, solutions and environmental impact	Congestion on the main motorways and urban roads High house prices Drinking and domestic water shortages Government is proposing to build many new houses but there is much local opposition Demand for more houses and factories creates pressure on existing land There are dangers in the highly volatile nature of the high-tech industries

Fig 13.6 The M4 corridor

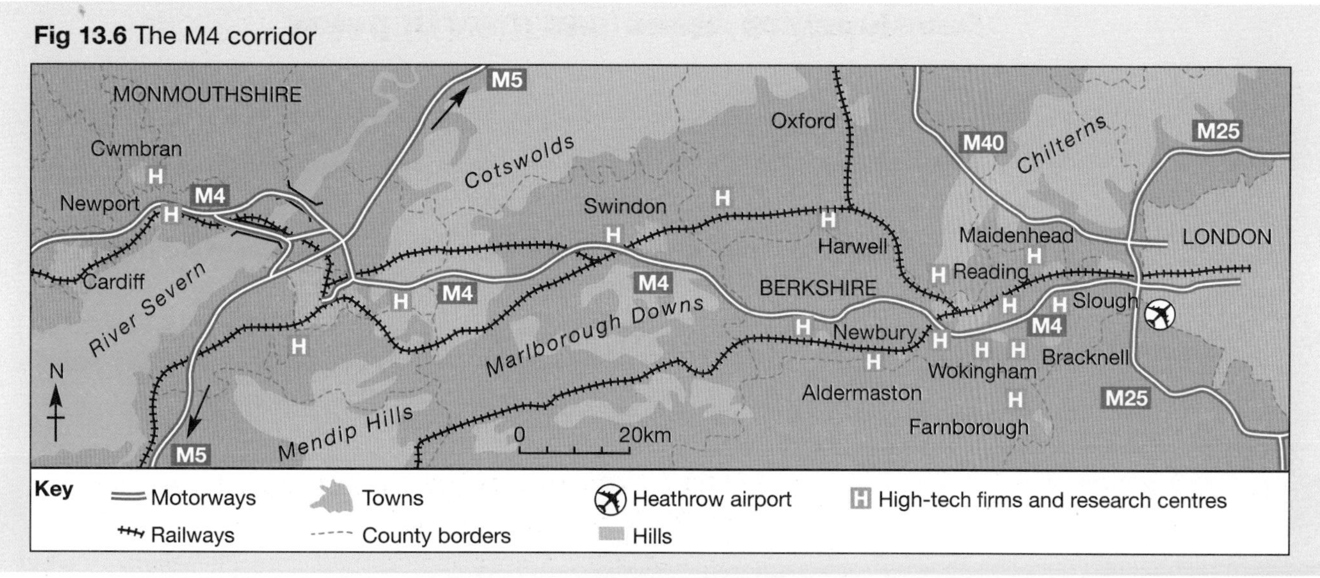

New sites for industry

AQA A AQA B

The globalisation of industry

This new term refers to the large Trans National Companies (TNCs) that operate in many countries. They use labour, raw materials and parts from many countries with the aim of producing goods people want at the lowest cost for the TNC. They sell their goods all over the world. The Ford Motor Company is one of these TNCs and its cars can be seen all over the world, sometimes under different names and with small changes in design. They have factories producing parts in at least 20 countries and the headquarters is in the USA. Many other companies, large and small, have followed this model and find the cheapest source for the production of their goods. Much of the clothing and footwear found on the British high street is made in countries in South-East Asia. The key point is the cost of production and especially the cost of labour.

The decline of manufacturing industry in MEDCs

AQA A AQA B

Traditional heavy industries such as iron and steel making have declined in many MEDCs. This is because:

- raw materials have become exhausted, e.g. iron ore in South Wales
- markets have moved to other countries, e.g. shipbuilding from Clydeside to the Far East
- competition from countries with cheaper labour, e.g. textiles from Lancashire to India and China
- older companies did not keep up to date by investing in new technology, e.g. car manufacture in the West Midlands.

The closure of **traditional industries** can have devastating effects upon an area. Unemployment spreads from the plant that closes to affect its suppliers. People in the area have less to spend so the local retail and entertainment industries suffer. People may not be able to continue their mortgage payments and house prices fall.

This was the case in the valleys of South Wales as coalmines, steelworks and associated industries closed or cut back.

Development areas and growth poles

There are a number of reasons why some governments recognise the need to develop industry including:

- the regeneration of an area that has seen the decline of its industry and increased unemployment, e.g. the decline of iron and steel and pottery in Stoke on Trent, the closure of coal mines in South Wales, the reduction of shipbuilding on Tyneside and in Belfast
- to encourage the setting up of industry in an area for the first time, e.g. growth poles in Brazil to attract people away from the crowded areas around Rio de Janeiro and São Paulo or rural development grants in remoter parts of Wales and Scotland.

Governments can encourage industry by establishing a **development area** where new companies can obtain benefits including:

- grants for the establishment of new factories
- ready-built premises with low rents
- small premises to allow the growth of new companies
- good road links to motorways and railway container depots
- advertising to promote the advantages of the area.

Enterprise Zones

These are special zones within areas of high unemployment and industrial decline where taxes are reduced and planning restrictions relaxed for ten years.

A case study of London Docklands would be useful here.

Urban Development Corporation

Following the success of Development Corporations in the building of new towns, the same strategy has been adopted for the redevelopment of some urban industrial areas. In this strategy, a piece of land is taken away from local government and a Development Corporation established. The Corporation has the responsibility, with the help of government and private finance, of redeveloping the area. This strategy was used to redevelop London Docklands. This derelict site of redundant docks has been transformed into a centre of employment with new office blocks and housing. Although this has created employment much of it has been unsuitable for the people who were living in the area.

Despite their success and popularity, government policy is to restrict further developments of these centres because they damage town centres, encourage car use that is environmentally damaging and use greenfield land.

The changing location of the retail industry

The retail industry includes shops and was traditionally located in the **Central Business District (CBD)**. In addition, shops were found in suburban high streets and local centres. Since the 1980s the UK has followed the pattern set in the United States with the retail industry moving to the edge of the urban area. The types of **out-of-town** shopping centre to be seen include:

- superstores containing a large supermarket or hypermarket and frequently a petrol filling station and car wash

- retail parks for large single-storey outlets for garden supplies, furniture, carpets, toys, pet food and accessories, DIY stores, sports goods and clothing
- regional shopping centres which consist of an undercover shopping mall, with two or three major chain stores acting as anchors and up to 200 other shops, cinema screens, restaurants and huge car parks. Some of these centres are served by their own light railway
- outlet villages such those at Bicester in Oxfordshire, Colne in Lancashire and the one occupying the former railway works at Swindon where chain stores sell discounted lines, often from the previous season's fashions.

These centres have been a great success. They provide plenty of car parking, good access most of the time, a range of shops often in warm and dry malls, safety as they are patrolled by security guards and competitive pricing. They do cause some problems at peak times when tailbacks can block surrounding roads and motorway slip roads.

The development of **out-of-town** centres has caused problems for town centres including:

- the loss of large shops has made the centre less attractive to shoppers
- smaller local shops have closed as fewer shoppers visit the centre
- vacant premises are often filled with less attractive discount and charity shops
- out-of-town centres are less accessible to older people and those without cars.

PROGRESS CHECK

1. List the factors encouraging high-tech industry along the M4 corridor.
2. Why are manufacturing industries declining in MEDCs?
3. Describe how the location of the retail industry has changed in recent years.

1. See page 168 2. See page 170 3. See pages 170–1

Chapter 14

Development, trade and aid

After studying this section you should be able to:

- ***describe different levels of development***
- ***understand plans to aid development***
- ***describe the main patterns of world trade***
- ***know the different types of aid***
- ***understand strategies to achieve sustainability***

Every country strives for development so that it can improve its economy and raise the quality of life of the people living there.

Development is the use of resources and technology to create wealth with which to improve the quality of life.

Different types of development

Here are three examples of development from three countries.

Development in Brazil

In places, the Amazon rainforest has been cleared for housing, agriculture and the Trans-Amazonian Highway as the country develops its timber and mineral ores. This development is raising the quality of lives of the people living there.

Development in India

The Green Revolution has been applied to agriculture in the Punjab. Income and living standards have improved for many people.

Development in the Caribbean

Tourism has been introduced to replace declining sugar cane and banana industries. Beach resorts, roads and power supplies have been provided and the jobs available are improving people's lives.

What is development?

AQA A AQA B AQA C

Development is the process of change which improves the well-being of a society in terms of material wealth and quality of life. Development includes:

- better food supply
- decreased infant mortality
- secure employment
- access to education
- warm dry housing
- improved health care
- longer life expectancy
- better working conditions
- security in old age
- water supply and sanitation.

Be able to list these 'quality of life' features. If you are asked how the quality of life in an area can be improved you can use these as headings.

There are four kinds of development:

- **economic development** including greater income and wealth through industrial growth
- **social development** with better standards of living, access to education, health, housing and leisure
- **environmental development** bringing improvements and restoration of the natural environment
- **political development** with progress to effective representative government.

Measuring the level of development

AQA A AQA C

There are number of measures of a country's level of development including:

- the changing percentage of labour in primary, secondary and tertiary jobs
- birth and death rates and rate of population growth
- infant mortality and life expectancy at birth
- adult literacy (being able to read and write) and access to secondary education
- **Gross National Product (GNP)**
- health measures such as people per doctor and access to clean water
- **Human Development Index**, created by the United Nations in 1990 to provide international comparisons. The index varies from 0 to 1 (most developed), and includes life expectancy, literacy, years in education and income per person.

KEY POINT

Gross National Product is the total value of all goods and services produced by a country in one year, divided by the population total to give an average amount per head.

You may be presented with a table of data about different levels of development. Read the data and units used carefully. If you have to make comparisons they can be between rows or columns.

Attempts to measure development have been criticised. For example, GNP does not include subsistence production for the person's own use. In addition, national figures may not be accurate and could be distorted by corrupt governments. Within a country there will be variations from the average national figure, e.g. between northern and southern Italy, south-east Brazil and Amazonia, and between cities and rural areas.

Table 14.1 Applying the measures to selected countries

Country	Birth rate per 1000	Death rate per 1000	Life expectancy (years)	Adult literacy rate (%)	People per doctor	Trade balance (million US$)	GNP per head (US$)
Ethiopia	48	18	47	33	38 000	–683	100
Brazil	25	7	66	82	729	7 561	3 020
Japan	10	8	79	99	613	121 825	31 450
UK	13	11	77	99	623	22 183	19 950

The main **development characteristics** of LEDCs are:

- birth rates are high and death rates are falling which leads to population increase and a strain on resources
- high infant mortality rates because of poor health services, inadequate diet and poor housing mean people have more children in the hope that some will survive
- lack of protection against killer diseases such as malaria and AIDS
- lack of finance to provide enough homes and remove some of the squatter camps/shanty towns
- increases in life expectancy increase the non-economically active population and puts a further strain on resources
- low levels of literacy, as education is not available (especially for females) or is expensive, leads to unskilled workforce
- lack of food or poor quality food leads to low work output
- shortage of clean water allows spread of diseases such as cholera, dysentery and diarrhoea.

Case studies of development issues

Water supply

If water supply is poor, people are not able to obtain clean drinking water or to have access to adequate sanitation. Poor water supply leads to poor health and disease. In Calcutta (Kolkata is its modern name) many people share a single cold water tap. In the bustees (shanty towns) toilets are almost non-existent and excrement is collected in plastic bags. The worn out sewers leak and contaminate drinking water supplies. As a result disease is rife.

Food supply

Despite food surpluses and a growing problem of obesity in many MEDCs there are still parts of the world where the people are underfed. In African countries, such as Chad, Ethiopia and the Sudan, up to two in every five people may not have sufficient food. There is not enough food in these countries because of problems such as war, political instability, debt and drought. Children are very vulnerable to the diseases marasmus and kwashiorkor if they suffer from malnutrition (lack of food or poor diet).

Influences on development

There are various factors involved in a country's **development**:

- political influences including political system, colonial influences and friendly nations
- physical environment including rainfall and water supply, natural hazards
- economic influences such as industrial structure, mineral wealth, agricultural potential.

Solutions to the problems of development

The solutions to the problems of development include:

- trade and aid
- sustainable development
- development projects in LEDCs
- appropriate technology
- role of the TNCs
- regional development plans.

1. How could you measure the difference in development between an LEDC and an MEDC?
2. Why was the Human Development Index invented?
3. Identify some solutions to the problems of underdevelopment in LEDCs.

1. See page 177 2. See page 177 3. See page 179

Trade and aid

  AQA C

Fig 14.2 The North–South development divide

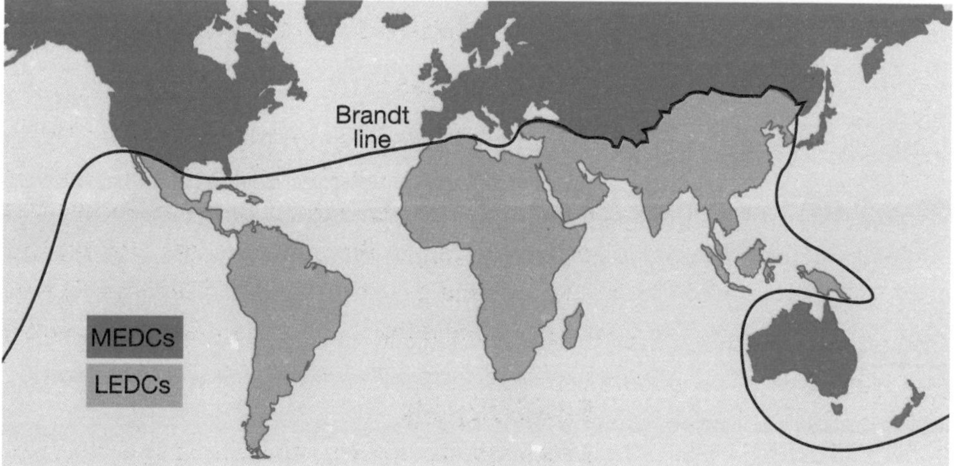

Fig 14.2 shows a map of the world divided into two areas. The map was drawn up by an international committee chaired by the former German Chancellor, **Willy Brandt** in 1980. To the north of the line are the developed richer MEDCs while to the south are the 'Third World' LEDCs. Since 1980 the gap between many MEDCs and LEDCs has become larger although some LEDCs such as Brazil, Malaysia, Indonesia and South Korea have developed considerably.

Trade

AQA A AQA B AQA C

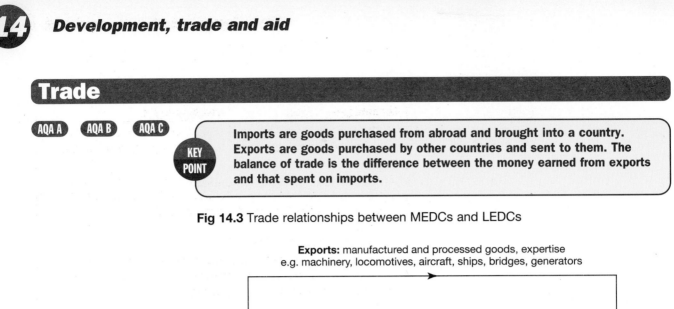

KEY POINT **Imports are goods purchased from abroad and brought into a country. Exports are goods purchased by other countries and sent to them. The balance of trade is the difference between the money earned from exports and that spent on imports.**

Fig 14.3 Trade relationships between MEDCs and LEDCs

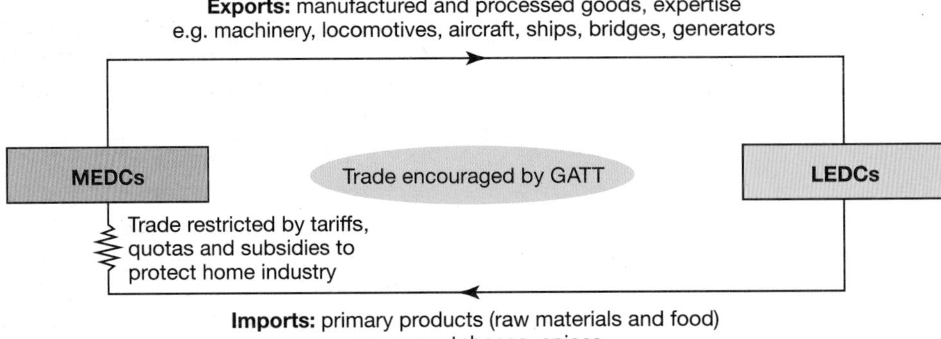

Pattern of trade

Trade is the movement of goods and services between producers and consumers. Governments and companies aim to purchase the cheapest goods available for a given quality. However, their choice also depends on trade agreements and other political factors. Japan, the EU and USA account for 60% of the world's trade.

Fig 14.4 The pattern of world trade

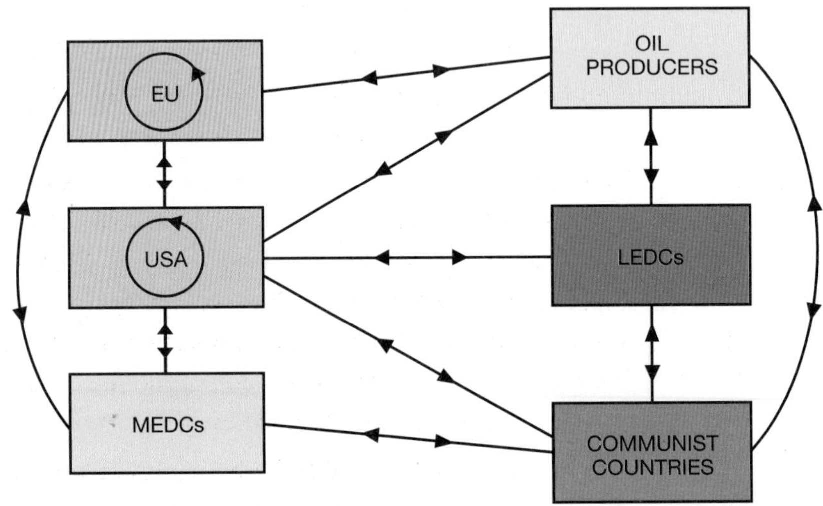

The advantages and disadvantages of trade

	Advantages	Disadvantages
Socio-economic	LEDCs gain foreign exchange by selling goods to MEDCs. MEDCs move factories to LEDCs to use cheaper labour. This allows LEDCs to develop. Goods produced are exported to MEDCs. MEDCs gain cheap imports of raw materials and export manufactured goods. LEDCs have a ready market for their goods.	LEDCs remain dependent on MEDCs to buy their products. Some emerging LEDCs dump cheap goods in MEDCs and this may cause unemployment There are dangers of unequal financial outcomes, e.g. trade deficits.
Environmental	LEDCs learn of ways to sustain and develop their resources without damaging their environment. Money from trade can be used to improve the environment.	MEDCs exploit resources in LEDCs and, with little sustainable provision, damage fragile environments. Factories in LEDCs producing goods for trade with MEDCs may give less attention to health and safety.
Political	MEDCs help governments in LEDCs to improve stability. MEDCs can exert pressure on LEDC government policies, e.g. in humanitarian and environmental matters. Trade with an MEDC may mean more Aid.	LEDC governments become dependent on orders from MEDCs. MEDCs may place conditions on LEDC governments if they wish to trade.

Trade and interdependence

Every country trades with other countries. Trade is important for the development of countries. When countries trade with each other they are said to be interdependent.

For example, Japan has a large population and is economically well developed. At the same time Japan has little flat land, few natural resources and limited energy sources. Japan imports machinery, food, oil, chemicals and textiles. These come mainly from the USA, Pacific Rim countries, China and the EU. In return, Japan exports electronic and other machinery, cars, chemicals and precision instruments to these trading partners. In an average year the value of Japan's exports is greater than its imports.

In contrast, Kenya is an LEDC with areas of rich soil where crops such as tea, coffee and fruit can be grown for export. Kenya has little manufacturing industry and has to import machinery, oil, cars, iron and steel and plastic to meet its needs. Over 40% of Kenya's trade is with the EU, with the UK its biggest single market. Each year the value of its imports is greater than its exports. Recently Japan and Kenya have become interdependent. Japan invests in Kenya and in return Kenya buys goods from Japan.

Fair Trade

A useful class discussion on how to make trade more fair would be worthwhile.

In the past powerful MEDCs have exploited LEDCs. They have purchased raw materials at low prices from LEDCs and sold expensive machinery to them. It has left the people of the LEDCs with low wages, little purchasing power and a poor quality of life. Today some organisations in MEDCs have recognised how unfair this trade can be for LEDCs. They are paying a 'fair' price for the goods they buy from the LEDCs. This allows people in LEDCs to have more money with which to buy the goods manufactured in MEDCs.

Aid

Why is aid needed?

Aid is the movement of resources from an MEDC to an LEDC and can include money, equipment, food, training, expertise or loans. The United Nations encourages countries to spend 0.7% of their GNP on aid but few do. Aid is intended to help LEDCs to continue their development and improve the quality of life of the people there.

Questions on aid are not popular with candidates. You should draw up a table of the advantages and disadvantages of the different types of aid for the giver and receiver.

Type of aid	Advantages	Disadvantages
Emergency (short-term), e.g. to areas hit by earthquakes, floods, hurricanes, tsunami, drought and volcanic eruptions.	Immediately provides food, clothing, shelter and medical assistance. Delivered directly to those in need with less chance of corruption.	Creates dependency. Treats the problem rather than the cause.
Political (governmental or conditional or tied), often bilateral when a richer country donates money or goods to a country in need but with strings attached.	Training and education available in MEDCs for people from LEDCs. Aids stability in LEDCs. Raises standard of living, educational level.	LEDCs has to buy from 'partner' MEDCs, e.g. arms, manufactured goods. Often has to be spent on prestige projects, and contractors have to come from MEDC.
Charitable (voluntary), e.g. non-governmental organisations (NGOs) such as Oxfam, Save the Children Fund and Comic Relief collect money in MEDCs to help people in LEDCs.	Money goes directly to the people in need. Also used for emergencies. Linked to low cost self-help schemes.	Many demands on charitable funds. Lottery prizes attract funds away from charities. Dangerous conditions in some unstable countries.
Long term (sustainable) where organisations commit for a long term programme, e.g. the Intermediate Technology Development Group.	Develops local skills to run and maintain equipment. Uses local raw materials in a sustainable way. Trains local people.	Economic activity remains at low level. Trained local people migrate abroad for higher wages.
International organisations (multilateral), e.g. the World Bank, United Nations, and the International Monetary Fund	Helps to train local people, e.g. farmers, and increases food production. Massive resources to help with large problems, e.g. disease.	May not be available to countries with hostile regimes. Danger of LEDC becoming dependent on the aid.

Appropriate technology and sustainable development

AQA A  AQA B AQA C

KEY POINT **Appropriate technology and sustainable development are alternative ways to reduce the gap between rich and poor.**

Appropriate technology

Some aid has been criticised as being too high-tech for LEDCs. The local people lack the expertise to maintain the project and have difficulty sustaining it. Spare parts are expensive or not available in the LEDC. Appropriate technology projects are low-tech, cheaper, use local materials and there is less to go wrong. They leave local people in control and there is less bureaucracy. Appropriate technology schemes often make use of the abundant local labour supply; products are cheap and local people can afford them.

E.F. Schumacher used the phrase 'small is beautiful' in 1973 for this type of development.

Examples of appropriate technology projects include:

- installing simple bamboo water pumps in villages
- providing bicycles for farmers to get goods to market
- using local streams to generate small amounts of hydro-electricity
- collecting rainwater in large clay pots
- using cement and chicken wire to reinforce walls and roofs.

Sustainable development

This refers to development which not only meets people's needs today but also those of future generations. We are moving towards **unsustainable development** as people consume more and more energy in using their mobile phones, the Internet, cars, air travel and air conditioning. If population and energy consumption continue to increase, the future of resources and the environment becomes more uncertain.

In 1992, the United Nations produced Agenda 21 with the following key points.

- Sustainable development is a local and a global concern involving local people and national governments.
- Local people and national governments are stewards of the Earth and should study how they consume resources and plan for a sustainable future.

The seven **key concepts** of sustainable development are:

The slogan for Agenda 21 is 'think globally and act locally'.

- **Interdependence**: people, the economy and the environment are interdependent at a range of scales from the local to global.
- **Citizenship and stewardship**: people have rights and responsibilities including participation and co-operation.
- **The future**: responsibility extends to protecting the needs and rights of future generations.
- **Diversity**: recognition and respect for the cultural, social, economic and biological diversity of our environment.
- **Equity**: acknowledgement and respect for the quality of life, equity and justice for all people.
- **Sustainable change**: plan and monitor change to ensure sustainability of developments and carrying capacity.
- **Uncertainty**: understanding of the uncertainty of outcome and taking appropriate precautions.

Examples of sustainable development

Commercial fishing

Sea fish stocks are being used up faster than fish are breeding. The European Union has imposed quotas, regulations on net size, closed seasons and exclusion zones to conserve stock. This has proved a tough policy for politicians, fishermen and consumers alike. However, it is vital if fishing is to continue in the long term.

Sustainable development is a key feature of every GCSE Geography syllabus. You must understand the term and know of examples in the UK and nationally.

Forests

Forestry provides a source of income and employment for some countries. It is very difficult to stop forest clearance and deny land to poor landless peasants who live near to forests, particularly tropical rainforests. Their governments may need the foreign exchange to improve agriculture or education and there is demand worldwide for their hardwood timber. Sustainable forestry allows some mature trees to be removed without decreasing biodiversity. The trees are cut and removed with care and new trees planted. This is hard to supervise in some of the more remote forests.

Building on brownfield land

The UK needs more houses and space for commerce and industry. Local authorities have been encouraged to develop derelict sites inside urban areas. Old factories, coal yards, railway sidings, gas works and military sites are being built on. This brownfield land may need clearing, access is not always easy and the local area not all that desirable. Developers and house buyers prefer housing on greenfield sites but building inside the city can regenerate it, cut travelling, reduce the need for new roads, preserve agricultural land and allow urban communities to be regenerated.

1. Why do some countries form trading agreements with other countries?
2. Identify the different types of aid.
3. What do you understand by the terms 'appropriate technology' and 'sustainable development'?

1. See page 177 2. See page 178 3. See page 179

Chapter 15

Tourism

LEARNING SUMMARY

After studying this section, you should be able to:

- ***define tourism and describe the ways in which it is changing***
- ***describe the growth and characteristics of tourism in LEDCs and MEDCs***
- ***understand the factors contributing to changes in tourism***
- ***recognise the impacts that tourism is having on places***

Leisure and tourism

Recreation becomes tourism when the person spends at least one night away from home.

Many people in MEDCs have free time after their day at work. After completing their duties at home they have time for leisure activities. In addition most people in MEDCs have paid holiday periods during the year. These trends are also appearing for people in paid employment in many LEDCs. People use this leisure time for recreation and tourism. Recreation is leisure time that lasts less than 24 hours. Recreation includes entertainment, skill improvement, rest and relaxation.

Tourism is a major earner and source of employment in the UK and many other countries.

What is tourism?

Tourism is about the increasing voluntary movement of people, for a limited number of days (temporary), from their place of residence to visit another place. It involves the places they pass through, those who make their trip possible, and people living at their destination.

Tourism is a tertiary or service industry that provides relaxation and enjoyment for one group of people and is a source of income, through a wide range of jobs, for others. Some of the jobs provided will be directly related to the tourists, e.g. waiters and guides, while others, such as doctors and shopkeepers, are less directly involved.

People become tourists for a number of reasons including:

- recreation
- to play or attend a sporting event
- to visit friends
- for adventure
- to celebrate
- to meet health needs
- to experience foreign culture
- to escape daily routine
- for prestige or status
- for education.

Remember that tourism involves an overnight stay and should not be confused with recreation.

Domestic tourists visit places in their own country while international tourists visit other countries.

The growth of tourism

AQA A AQA B AQA C

Tourism is the **largest** and **fastest-growing industry** in the world today. It is estimated that 10% of the people in paid employment work in the tourist industry and that there are over 700 million tourist arrivals worldwide (Table 15.1).

Table 15.1 International tourist arrivals worldwide and by region

Region	1985 world share (%)	1999 world share (%)	1985-1999 annual growth (%)
World	100	100	5.5
Americas	20	20	5.6
Africa	3	4	7.5
Asia	10	15	10.0
Europe	65	59	4.5
Middle East	2	2	2.6

Many factors account for the growth of tourism in some places. They can be divided into two groups (Table 15.2).

Table 15.2 Demand and supply factors in tourism

Demand factors: Factors increasing the number of people wanting to be tourists	Supply factors: Factors encouraging the development of tourism
• People have increased amounts of income available for holidays (greater affluence) • Jobs provide paid holidays • The length of the working week has been reduced • Travel is easier • Cheaper package holidays are available • Advertising increases awareness • Hectic urban lifestyles • Modern telecommunications, including telephones, fax machines and the Internet make international booking more efficient • Early retirement and a longer life expectancy mean that more people over 60 are tourists	• Attractive climates, including warm to hot temperatures, little rain, or snow in winter for skiing • Culture including buildings, museums, art galleries and festivals, distinctive food, traditional music • The combination of sun, sea and sand • Safety and stable government • Special festivals or sporting events • Nostalgia or adventure • Scenery such as mountains, gorges, waterfalls, coasts and caves • Ecology including interesting trees, plants, birds and marine life • Famous sites of historic events • Sports facilities and nightlife

Types of tourist break

Be ready to account for the growth of tourism in some places but not in others.

Be ready to suggest how each of these could be applied to places you know.

There are many options available to tourists including:

- **package holidays** which include flight, accommodation, food and a guide
- **weekend breaks** at hotels mainly used by business people during the week
- **family holidays** designed to provide particularly for children with organised events, child-minding facilities and evening entertainment
- **age-specific holidays** such as those for the over-55s or younger people between 18 and 30
- **self-catering** in an apartment, small cottage or house where holiday makers cook most of their own meals
- **fly-drive** which involve a flight and rental car and where hotels and itinerary may be arranged in advance
- **individual or independent group**, may be on a low budget or for an extended period such as backpacking across an area
- **specialist breaks** to develop a skill including sailing, painting, photography, gardening.

1. Suggest difficulties you might have in defining and measuring the growth of tourism.
2. Why does tourism develop in some places and not in others?
3. Why do both busy and quiet seasons pose problems for holiday resorts?

1. Not all travellers are tourists. Some people mix business and pleasure; no official records are kept. 2. See page 182 3. Pressure in busy seasons and unemployment and lack of income in quiet seasons.

The advantages and disadvantages of tourism in MEDCs

AQA A AQA B AQA C

Case study: National Parks

National Parks are attractive areas that are managed for the enjoyment of tourists and the benefit of local people.

There are National Parks in the USA (where the idea began), the UK, Malaysia and several African countries. A National Park is a large area of attractive countryside, often including small towns and villages, which is protected by law. In the USA, the National Parks tend to be wilderness areas.

National Parks in England and Wales

In 1949 the government passed the National Parks and Access to the Countryside Act to identify National Parks. In England and Wales there are ten National Parks of which the Peak District was the first. In addition the New Forest and South Downs are becoming National Parks and the Norfolk Broads is a Specially Protected Area.

Fig 15.1 National Parks in England and Wales

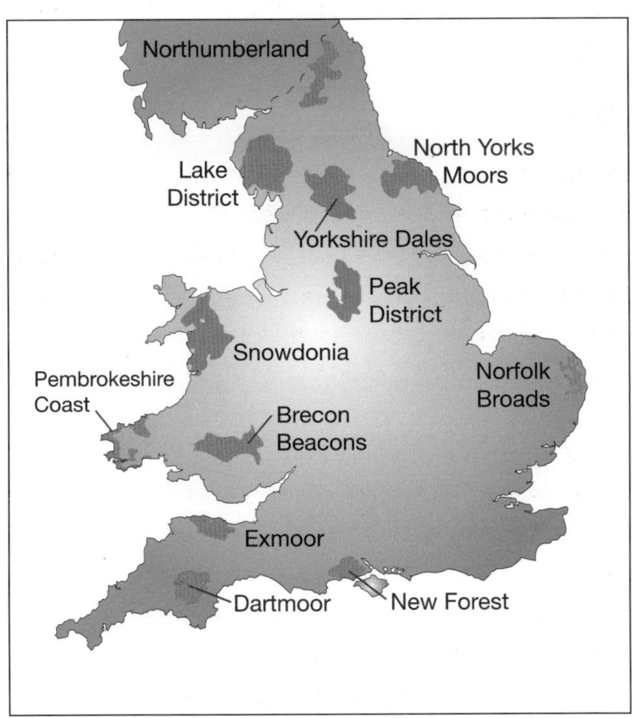

Within **National Parks** the beauty of the landscape is preserved, the public is given access to much of the countryside; wildlife and important buildings are protected and traditional farming methods encouraged. The largest landowners are the National Trust, Forestry Commission and Ministry of Defence although well over 90% of the land is privately owned. A **National Parks Authority (NPA)** administers each National Park. Each Authority manages an area where economic activity continues, e.g. agriculture, forestry and quarrying, where local communities find their livelihood and large numbers of visitors arrive to walk, cycle, climb, sail and visit places of interest.

National Parks contain beautiful upland scenery with hills, moors, lakes, and forests and, in nearly all cases, spectacular coasts. The growth of the leisure industry together with mass car ownership means that more people are visiting the National Parks. These changes have brought both opportunities and problems for these areas.

Get to know a case study of one National Park and to name examples of places in it.

Opportunities

- The growing number of visitors bring money into the area.
- Visitors provide jobs and allow shops and other services to thrive.
- Roads and railways are maintained.
- Communities remain alive as people stay to work locally.
- A rich cultural life survives as audiences are supplemented by visitors on holiday.

Problems

- Traffic congestion occurs as poor local roads become crowded at peak times.
- Car parks fill up and grass verges are damaged by illegal parking.
- Footpaths are eroded by the large numbers of walkers.
- Bridleways become muddy with the increased number of mountain bikers and horse riders.
- Some settlements and sites become over-crowded to the point where their original attractiveness is threatened (these are called '**honeypot**' sites).
- House prices rise out of the reach of local young people as second home buyers move in from the large urban areas.
- Local people convert houses into holiday cottages and reduce the number available to local people.
- Farmers have their working land invaded by visitors causing damage to fences, crops and animals.

Managing the problems

Methods used to manage the problem in National Parks include:

- providing park-and-ride schemes on the edges of the sensitive areas
- reinforcing footpaths and providing alternative routes
- focusing demand on one or two honeypot sites and accepting they will be sacrificed for the sake of other areas
- ensuring that affordable new housing is built for local people
- demanding that quarrying is landscaped during and after use.

The advantages and disadvantages of tourism in LEDCs

AQA B AQA C

KEY POINT **Many LEDCs are attractive destinations for rich tourists but lack stable management.**

The growth of tourism

LEDCs such as the Caribbean, China, Kenya, Zimbabwe, Thailand, Malaysia, Mexico and Egypt have well-established tourist industries. They have developed their tourist industry as a way of improving their economies and the quality of life for the people living there. The reasons for the growth of tourism in LEDCs include:

- government projects to build new airports, roads and hotels have been supported by investment from MEDCs

- costs are lower in LEDCs and they become cheap places for people to visit
- companies promoting tourism have been able to grow by advertising holidays in LEDCs
- traditional holiday resorts, such as the Mediterranean, have become crowded with noise, pollution and high costs
- people are looking for new experiences including the exotic
- good weather is more reliable in some LEDCs
- long-haul flights have become cheaper as larger aircraft carry more passengers directly to their destinations.

Many LEDCs see tourism as a way of getting started on the path to development. They need to have stable government if they are to succeed in the long term.

For example, in Zimbabwe, tourists are able to fly to Harare where there are many good hotels. The weather is very predictable and good roads allow access to the countryside. The country has many interesting ancient sites such as the Great Zimbabwe ruins, the spectacular Victoria Falls, several impressive dams, such as that at Kariba, and interesting local traditions including sculpture. Costs are very low and high quality food plentiful. Recently political unrest has threatened this growth.

Tourism brings advantages and disadvantages to LEDCs

The **advantages or benefits** brought by income from tourism include:

- economic development can be promoted
- schools and hospitals can be built
- hotels provide jobs for people and a market for farmers' products
- employment is created in a wide range of industries including taxi driving, the making of gifts and souvenirs, travel guides and entertainment.

Notice how words can be interchanged. 'Benefits' and 'advantages' are positive effects. Disadvantages can include 'issues', 'problems', and 'conflicts'. Question setters select the right word to guide you in answering the question.

The **disadvantages, problems, issues and conflicts** include:

- planning controls are not strong in many LEDCs and uncontrolled development, especially of hotels, can make coastal areas and islands unattractive
- fishermen and other traditional activities can be displaced by coastal development
- many tourist activities are owned by foreign firms and much of their earnings leaves the LEDC
- farming may change to grow the types of food liked by tourists or food is imported to the disadvantage of local markets
- local culture may become modified to suit the daily shows for tourists
- crime, prostitution and drug use may increase
- tourists may have little respect for local traditions and religions
- the jobs created are casual and poorly paid
- valuable water resources are used for swimming pools and showers
- damage may be caused to the ecosystem through poaching or over-use of sites.

1. How many National Parks can you name in England and Wales?
2. List advantages and disadvantages of the growth of tourism for people living in National Parks.
3. Name some countries and the attractions they have for tourists.

1. See page 183 2. See page 184 3. France – Paris, wine, food, climate, coasts, countryside; China – Great Wall, Forbidden City, Terracotta Warriors

MEDC case study: tourism in the Mediterranean

The Mediterranean coast

The coastline and islands of the Mediterranean provide many tourist sites for people living in the cooler and less sunny parts of Europe including the UK, Scandinavia and Germany. The eastern coast of Spain, the French Riviera, and the southern coast of Greece, as well as islands such as Majorca, Cyprus and Crete, are the main tourist areas. The main holiday season is from April to October and during this period the population of the coastal area can double. This causes problems for local services such as water and sewerage whilst boosting employment in the hotel and other service industries.

The coast of the Mediterranean is some distance from the industrialised core of Europe. Before the growth of tourism the Mediterranean coast was part of the poorer periphery of Europe.

It can be helpful to use the terms 'core' and 'periphery' when describing distributions. Be sure of their definitions.

The core is the area of a country or region where economic activity is at its greatest. Population, industry and commerce are concentrated in the core. In contrast, the periphery is remote from the core and has low population density, high unemployment and out-migration.

There are both costs (disadvantages) and benefits from tourism for people living in the Mediterranean.

Costs of tourism

The costs include:

- pollution: coastal settlements and industrial sites, such as oil refineries and chemical plants, discharge storm water, sewage and waste into the almost landlocked and tideless sea
- poorly-planned building development including hotels, shops and apartments
- loss of farm land to building as the price of land rises
- congestion, despite the building of intrusive new roads and motorways, as both visitors and local people have access to cars
- increased noise from flights and night clubs as family holidays tend to take up 24 hours with their different activities
- the occupation of local shops and services with souvenir and gift shops, fast food restaurants and places of entertainment
- increased crime, alcoholism, drugs and bad behaviour as young people have 'a good time'.

Why do people from the economic core of Europe choose to visit the Mediterranean for holidays?

Benefits of tourism

The benefits include:

- a variety of jobs are provided in areas where farming and fishing or unemployment used to be the norm
- local people have access to a higher standard of living as tourists spend money
- better roads, hospitals and shops are provided to meet the needs of visitors
- local people are able to use the tourist attractions such as bars and night-clubs
- the tourists bring new ideas and interests to the area
- funds are made available to conserve and restore old buildings.

LEDC case study: tourism in Singapore

Singapore is a small crowded island on the tip of the Malaysian peninsula. It is located very close to the Equator. Singapore is a highly organised country that built its initial wealth upon the trade passing through it as a port. With the growth of global tourism, people in Singapore have set out to make it a tourist attraction despite its small crowded site.

The number of **tourists arriving** in Singapore has risen rapidly since 1990. Visitors, on arriving in Singapore while having their passports checked by the immigration authorities, are asked to fill in a form. The form asks them to state the purpose of their visit. Most of the visitors are from South-East Asia, Japan and Taiwan. The main purpose of their visit is for holidays and the average length of stay is just under a week. Quite a large number of visitors are in transit.

Singapore's attractions as a tourist destination

The main attractions of Singapore that contribute to the tourist industry are:

- the equatorial climate with sunny mornings and warm evenings
- flora and fauna in the two nature reserves
- bird watching
- coastal activities including sea and sand on the offshore islands
- cultural events and the arts and crafts associated with the Indian, Malay and Chinese communities
- the National Museum, Parliament building and Kanji War memorial are popular sites for visitors
- the annual shopping sale, the international Dragon Boat racing and the Festival of the Arts.

Sustainable tourism

AQA A AQA C

The growth of tourism threatens the very places the tourists visit. The erosion caused by people walking across a site and the moisture from their breath causes damage. The Taj Mahal, Stonehenge, the caves at Lascaux in the Dordogne, the Great Barrier Reef in Australia and the pyramids in Egypt are just five sites where access has been restricted. Summer and autumn visitors to quiet villages in the Cotswolds and Lake District find the peace they came to find destroyed by the large numbers of people. Even Sunday climbers find themselves queuing to use the chalk-marked holds as too many people arrive in Snowdonia, North Wales, for a mountain experience.

Sustainability is a very important word in new GCSE syllabuses. Know what it is and learn about sustainable tourism.

People managing tourist sites or planning new ones are looking at alternative ways in which to develop them. They want the tourism to bring the revenue but the attractions of the site to be sustained. Sustainable plans include:

- ecotourism
- honeypot sites and wilderness areas.

Ecotourists visit areas to learn about the people living there, the local culture and environment without causing damage. At the same time, ecotourists want the local people to benefit from the economic advantages that they bring.

Ecotourism (sometimes called green tourism)

Ecotourism is taking place in some areas. It involves training local guides to take tourists to a variety of locations so that no one place is visited by too many people. During the visit, the guide makes sure that no damage is caused and the visitors are able to enjoy the wildlife and vegetation around them. It also means teaching local people to respect and protect their own local environment. Some of them may be tempted to make quick money by opening conventional tourist attractions or selling land to developers. They are shown that ecotourism brings increased income whilst protecting and improving their way of life.

The advantages of this approach are:

- some of the income is used to protect the natural environment
- the quality of local people's lives is improved without their culture and way of life being destroyed
- local people are able to remain on their land as farmers as well as working in the tourist industry
- the area is conserved for the benefit of future tourists.

Case study of tourism benefitting the environment, local people and the country

Ways in which countries can maintain their tourist trade

Better management

- tighter building regulations and laws
- creating jobs for locals that protect scenery, wildlife and culture
- resistance to short-term proposals

Preservation of attractive sites

- planning restrictions
- greater understanding of local culture
- penalties for exploitation of local sites

Enhancing the environment

- encouraging ecotourism, small groups
- guides to point out need to protect threatened features
- promote specialist interests and more environmental education

Tourism in Keyna

The environment

- authorities protect the environment to preserve it for the tourists
- National Parks set up to protect and manage the environment
- hunting is banned
- tourists expected to keep to clearly marked tracks

Local people

- provides employment, e.g. drivers of safari journeys, building safari lodges, hotel staff
- local customs recognised as attractive to tourists and protected
- selling goods and services bring local people money

The country

- a major source of overseas income
- income used to provide services for local people

No one is certain that this plan will work. Contact with visitors changes the ways in which local people want to live. Even small numbers of visitors lead to changes, as they demand quality hotels, fast communication routes, air conditioning, familiar food and sewerage facilities.

Honeypot sites and wilderness areas

This strategy involves identifying some popular locations and allowing them to develop as honeypot sites. Bigger car parks are built, more public toilets provided, hotels and restaurants extended, cycle rides and footpaths marked, access roads improved and housing built. These popular sites become even more popular and very crowded in the summer. Local residents may complain but the shopkeepers, hotel and restaurant owners thrive.

Allowing these honeypot sites to attract so many people draws demand away from other sites and they are protected. Around the less popular sites, access is not improved and car parking is very limited. These areas remain for farming or moorland and, in the more remote parts, as wilderness. In this strategy, some popular sites are sacrificed to protect all the others

1. Describe the difference between the core and periphery of a country.
2. For a place you have studied, describe the costs and benefits it receives from its tourist industry.
3. What do you understand by ecotourism? How can it be developed in a place you have studied?

1. See page 186 2. See pages 186–7 3. See page 188

Ordnance Survey map

1:50 000 extract Keswick, Lake District

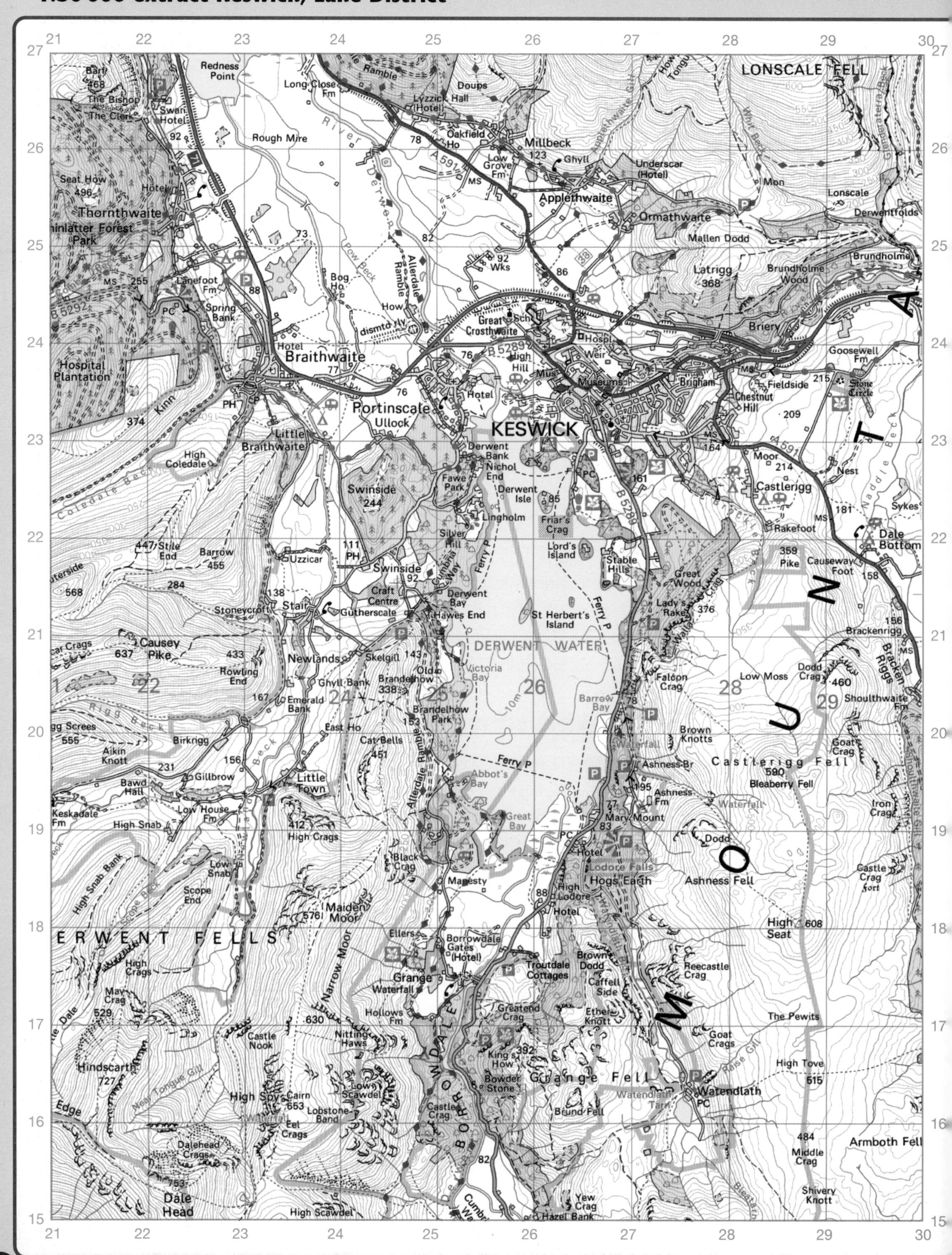

Ordnance Survey key

Communications

ROADS AND PATHS

Unfenced Dual carriageway A 470	Primary Route
Footbridge A 493	Main road
B 4518	Secondary road
A 855 B 885	Narrow road with passing places
Bridge	Road generally more than 4m wide
	Road generally less than 4m wide
	Other road, drive or track
	Path Gradient : steeper than 20% (1 in 5) 14% to 20% (1 in 7 to 1 in 5)
Gates Road tunnel	

PRIMARY ROUTES

These form a network of recommended through routes which complement the motorway system

PUBLIC RIGHTS OF WAY

-------------- Footpath

-------- Bridleway

-·-·-·-·- Road used as a public path

-+-+-+-+- Byway open to all traffic

OTHER PUBLIC ACCESS

• • • Other route with public access { not normally shown in urban areas

◆ ◆ National Trail, European Long Distance Path, Long Distance Route, selected Recreational Routes

● ● ● National/Regional Cycle Network

[4] National Cycle Network number

Tourist Information

TOURIST INFORMATION

⚑	Camp site	PC	Public convenience (in rural areas)
🚐	Caravan site		Selected places of tourist interest
❋	Garden	☎ ☎	Telephone, public / motoring organisation
⚑	Golf course or links		
i i	Information centre, all year / seasonal	☀	Viewpoint
🦆	Nature reserve	V	Visitor centre
P P&R	Parking / Park and ride, all year / seasonal	!	Walks / Trails
✕	Picnic site	▲	Youth hostel

HOW TO GIVE A NATIONAL GRID REFERENCE TO NEAREST 100 METRES

SAMPLE POINT: Goodcroft

1. Read letters identifying 100 000 metre square in which the point liesNY

2. FIRST QUOTE EASTINGS
Locate first VERTICAL grid line to LEFT of point and read LARGE figures labelling the line either in the top or bottom margin or on the line itself ...53
Estimate tenths from grid line to point ..4

3. AND THEN QUOTE NORTHINGS
Locate first HORIZONTAL grid line BELOW point and read LARGE figures labelling the line either in the left or right margin or on the line itself ..16
Estimate tenths from grid line to point..1

SAMPLE REFERENCE NY 534 161

For local referencing grid letters may be omitted

IGNORE the SMALLER figures of the grid number at the corner of the map. These are for finding the full coordinates. Use ONLY the LARGER figure of the grid number. EXAMPLE: 31⁷000m

General Information

LAND FEATURES

Electricity transmission line (pylons shown at standard spacing)	
> - - > - - > Pipe line (arrow indicates direction of flow)	
ruin Buildings	Cutting, embankment
	Quarry
Public building (selected)	Spoil heap, refuse tip or dump
Bus or coach station	Coniferous wood
Place of Worship { with tower	Non-coniferous wood
{ with spire, minaret or dome	Mixed wood
+ { without such additions	Orchard
○ Chimney or tower	Park or ornamental ground
Glass Structure	
H Heliport	Forestry Commission access land
△ Triangulation pillar	
Mast	National Trust-always open
Wind pump/wind generator	National Trust-limited access, observe local signs
Windmill with or without sails	
+ Graticule intersection at 5' intervals	National Trust for Scotland

ABBREVIATIONS

CH	Clubhouse	P	Post office
PC	Public convenience (in rural area)		
TH	Town Hall, Guildhall or equivalent	PH	Public house

ARCHAEOLOGICAL AND HISTORICAL INFORMATION

+ Site of monument	✗ Battlefield (with date)	VILLA	Roman
· ○ Stone monument	☆···· Visible earthwork	Castle	Non-Roman

HEIGHTS

—— 50 —— Contours are at 10 metres vertical interval

·144 Heights are to the nearest metre above mean sea level

ROCK FEATURES

Outcrop Cliff Scree

KILOMETRES 1 0 1 2 3

Scale 1: 50 000

N

1: 50 000 scale Second Series

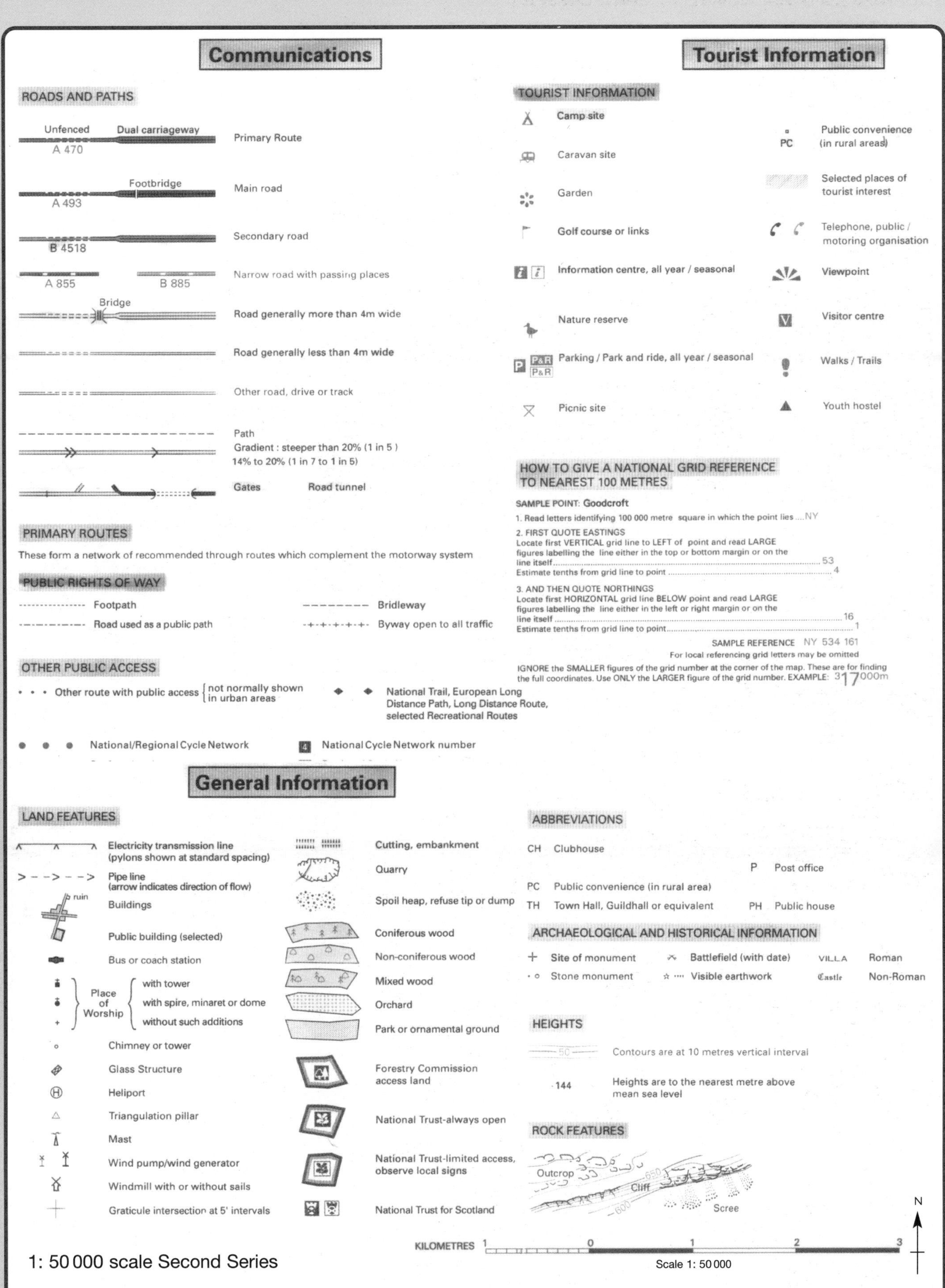

Ordnance Survey map questions

The questions on the Ordnance Survey map are compulsory. If you have not been practising your map skills you will not do very well in this paper. The following questions cover most of the skills you require but you need to practise on other 1:50 000 and 1:25 000 maps. Keys will be provided for all maps, but it is important that you are familiar with the OS keys at both scales. Use the latest maps – some new symbols have been introduced recently.

1. **(a)** What feature is shown marking the summit of High Spy GR 234162?

.... **[1]**

(b) If you take the footpath from High Spy to Maiden Moor 237182

(i) in which direction will you walk? ....

(ii) how far will you walk (in kilometres)? ....

(iii) what landform feature will you walk along? ....

(iv) what are the landforms shown just below the summit on the west side of High Spy? **[4]**

....

2. Start at the car park at GR 272212.

(a) What type of land use surrounds the car park? .... **[1]**

(b) If you take the footpath to Bleaberry Fell GR 285196 how far will you have to walk (in kilometres)? .... **[1]**

(c) Describe your walk referring to land use and how steep the climb will be. **[4]**

....

....

3. The figure shows part of a cross-section drawn between Bleaberry Fell GR 285196 and Cat Bells GR 244198.

(a) Complete the cross-section. **[3]**

(b) How steep is the gradient of the slope from the summit of Cat Bells (451m) to the lake (70m):
about 1:2?
about 1:7?
about 1:10? **[1]**

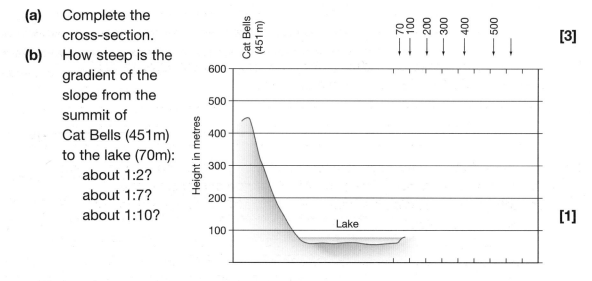

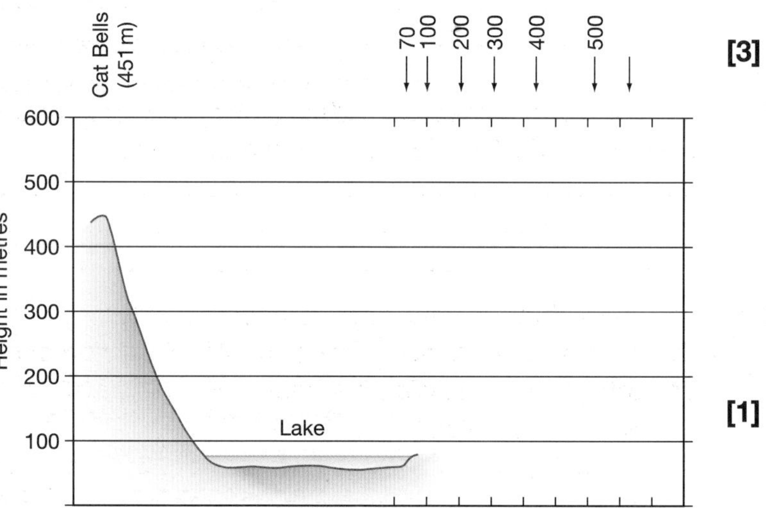

Ordnance Survey map questions

(c) Describe the shape of the valley shown on the cross-section. [3]

....

....

(d) Indicate on the cross-section **two** areas of woodland and **one** area of moorland. [3]

4. Refer to grid squares 2424/2524 **and** grid squares 2528/2628. Compare the valley and course of the River Derwent with that of Whit Beck. [4]

....

....

5. (a) What evidence is there in grid square 2523 that the area is popular with tourists?

.... [2]

(b) Give the six-figure map reference of the bus station in Keswick.

.... [2]

(c) What tourist feature is found at grid reference 292238? .... [1]

6. (a) Describe the pattern of roads shown on the map extract.

....

.... [3]

(b) What evidence is there on the map extract to suggest that the building of a railway to Keswick was difficult?

.... [2]

7. If you were travelling along the B5289 from Keswick to Barrow Bay car park (270204), in which direction would you be heading? .... [1]

8. Describe the pattern of settlement shown on the map extract.

....

....

.... [4]

9. (a) Describe two pieces of evidence shown on the map that suggest that Keswick had a problem with traffic congestion.

....

....

.... [4]

(b) The map shows many public footpaths including the Cumbrian Way along the west side of Derwent Water (2521). Describe and explain one advantage and one disadvantage this would have for people living on that side of the lake.

....

....

....

.... [6]

[Total 50 marks]

Sample GCSE question

1. (a) Study the map, which shows the global distribution of earthquakes.

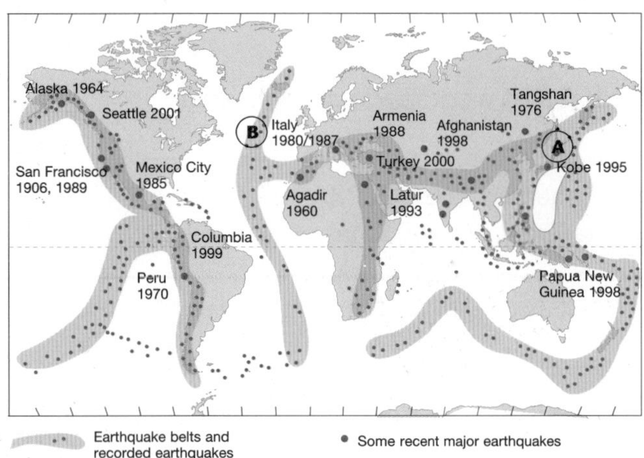

(i) Describe the pattern of earthquake activity shown on the map. **[2]**

Earthquakes occur in long narrow bands. ✓ *There is a dense ring around the Pacific Ocean.* ✓

> Develops the answer by locating a specific pattern.

(ii) Explain why earthquakes occurring at **A** are likely to be of a larger magnitude (i.e. stronger) than those occurring at **B**. **[4]**

Earthquakes at A are where two plates are converging. The oceanic plate is forced under the continental plate causing a lot of pressure and friction. ✓ *When the pressure is suddenly released it can cause a violent earthquake.* ✓

At B the plates are moving apart. Pressure is slowly released ✓ *and earthquakes are less violent.* ✓

> A good answer – carefully applies knowledge of plate margins to explain differences in magnitude of earthquakes.

(b) The figure shows a sketch cross-section of Soufriere Hills volcano, Montserrat (Caribbean).

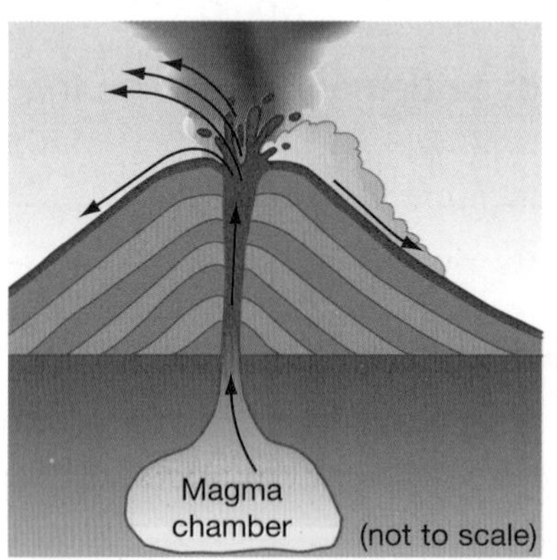

(i) Describe **three** types of material that may be ejected from the crater when a volcano erupts. **[3]**

1. Molten rock called lava ✓

2. Ash is thrown high into the air ✓

> Correct but read the question carefully – it asks for three types of material. A wasted mark here.

(ii) By referring to two different locations of volcanic activity, describe some positive and negative effects on human activity. **[6]**

In Montserrat ash and lava buried the capital city of Plymouth ✓ *and destroyed the airport and only hospital.* ✓ *Many people had to be evacuated from the island.* ✓ *In some places volcanic activity is used to make electricity for industry.* ✓

> Well developed first part but second example lacks a location and the effect is not developed.

[Total 15 marks]

Sample GCSE question

2. (a) Study the figure, which shows some landforms of glacial deposition.

(i) Name the features labelled **X**, **Y** and **Z.** **[3]**

X - drumlin ✓
Y - erratic ✓
Z - terminal moraine ✓

Uses the correct terms – no need for sentences.

Glacier retreating
Valley side
Meltwater stream
Solid rocks
X
Y
Z

(ii) Choose one of the depositional landforms you have labelled in (i).
Explain how it was formed. **[4]**

I will explain feature Y.
When ice moves it erodes large boulders and pebbles from the rocks below it by a plucking action. Glaciers are able to carry these boulders away as they move down the valley. ✓ When the ice melts it deposits and leaves all the material it is carrying on the surface below. ✓

Rocks moved away from their source like this are called erratics. ✓ Some erratics found in Norfolk have been moved from Norway. ✓

The answer describes a clear sequence and uses the correct terminology. It concludes with a good example, which supports the explanation.

(b) Study the map showing the National Parks in England and Wales.

Northumberland
Lake District
North Yorks Moors
Yorkshire Dales
Peak District
Snowdonia
Pembrokeshire Coast
Norfolk Broads
Brecon Beacons
Exmoor
Dartmoor
New Forest

(i) Give **two** reasons why the Lake District and Snowdonia National Parks are popular tourist venues. **[2]**

Both these national parks have mountain scenery, which attracts lots of tourists. ✓ Several large cities, such as Manchester, are not far from these parks and many people like to get away from the crowded cities at weekends. ✓

Two good points supported by an example.

(ii) Explain how tourism in a National Park such as the Lake District can have both positive and negative effects on people living locally. **[6]**

Tourism provides local employment in hotels and shops. ✓ Farmers are able to diversify. ✓ Large numbers of visitors cause traffic congestion ✓ in places such as Keswick. ✓

To gain six marks, points need to be developed. Three points are made here but only the negative one is developed.

[Total 15 marks]

Sample GCSE question

3.

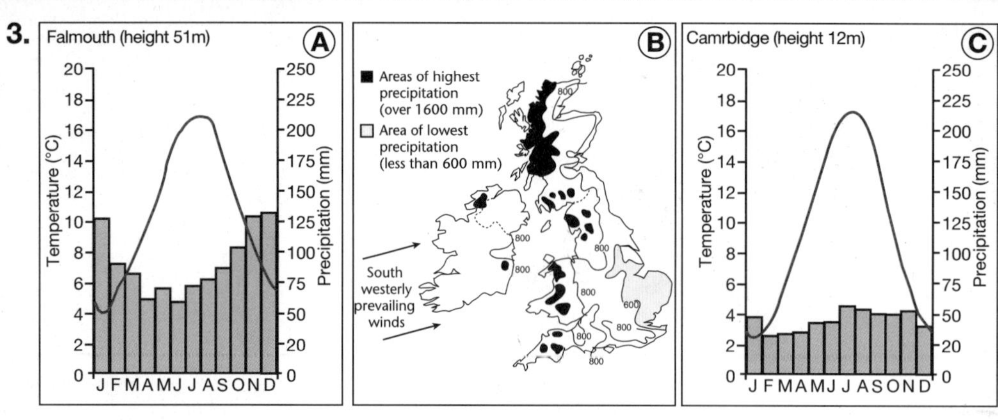

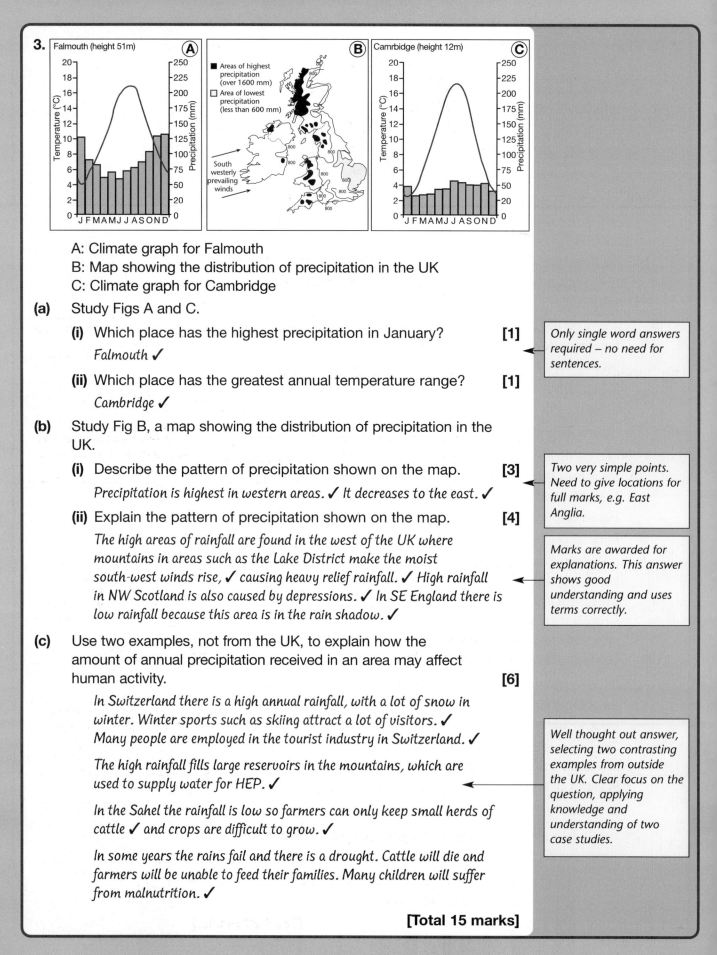

A: Climate graph for Falmouth
B: Map showing the distribution of precipitation in the UK
C: Climate graph for Cambridge

(a) Study Figs A and C.

(i) Which place has the highest precipitation in January? **[1]**

Falmouth ✓

(ii) Which place has the greatest annual temperature range? **[1]**

Cambridge ✓

> *Only single word answers required – no need for sentences.*

(b) Study Fig B, a map showing the distribution of precipitation in the UK.

(i) Describe the pattern of precipitation shown on the map. **[3]**

Precipitation is highest in western areas. ✓ It decreases to the east. ✓

> *Two very simple points. Need to give locations for full marks, e.g. East Anglia.*

(ii) Explain the pattern of precipitation shown on the map. **[4]**

The high areas of rainfall are found in the west of the UK where mountains in areas such as the Lake District make the moist south-west winds rise, ✓ causing heavy relief rainfall. ✓ High rainfall in NW Scotland is also caused by depressions. ✓ In SE England there is low rainfall because this area is in the rain shadow. ✓

> *Marks are awarded for explanations. This answer shows good understanding and uses terms correctly.*

(c) Use two examples, not from the UK, to explain how the amount of annual precipitation received in an area may affect human activity. **[6]**

In Switzerland there is a high annual rainfall, with a lot of snow in winter. Winter sports such as skiing attract a lot of visitors. ✓ Many people are employed in the tourist industry in Switzerland. ✓

The high rainfall fills large reservoirs in the mountains, which are used to supply water for HEP. ✓

In the Sahel the rainfall is low so farmers can only keep small herds of cattle ✓ and crops are difficult to grow. ✓

In some years the rains fail and there is a drought. Cattle will die and farmers will be unable to feed their families. Many children will suffer from malnutrition. ✓

> *Well thought out answer, selecting two contrasting examples from outside the UK. Clear focus on the question, applying knowledge and understanding of two case studies.*

[Total 15 marks]

Sample GCSE question

4.

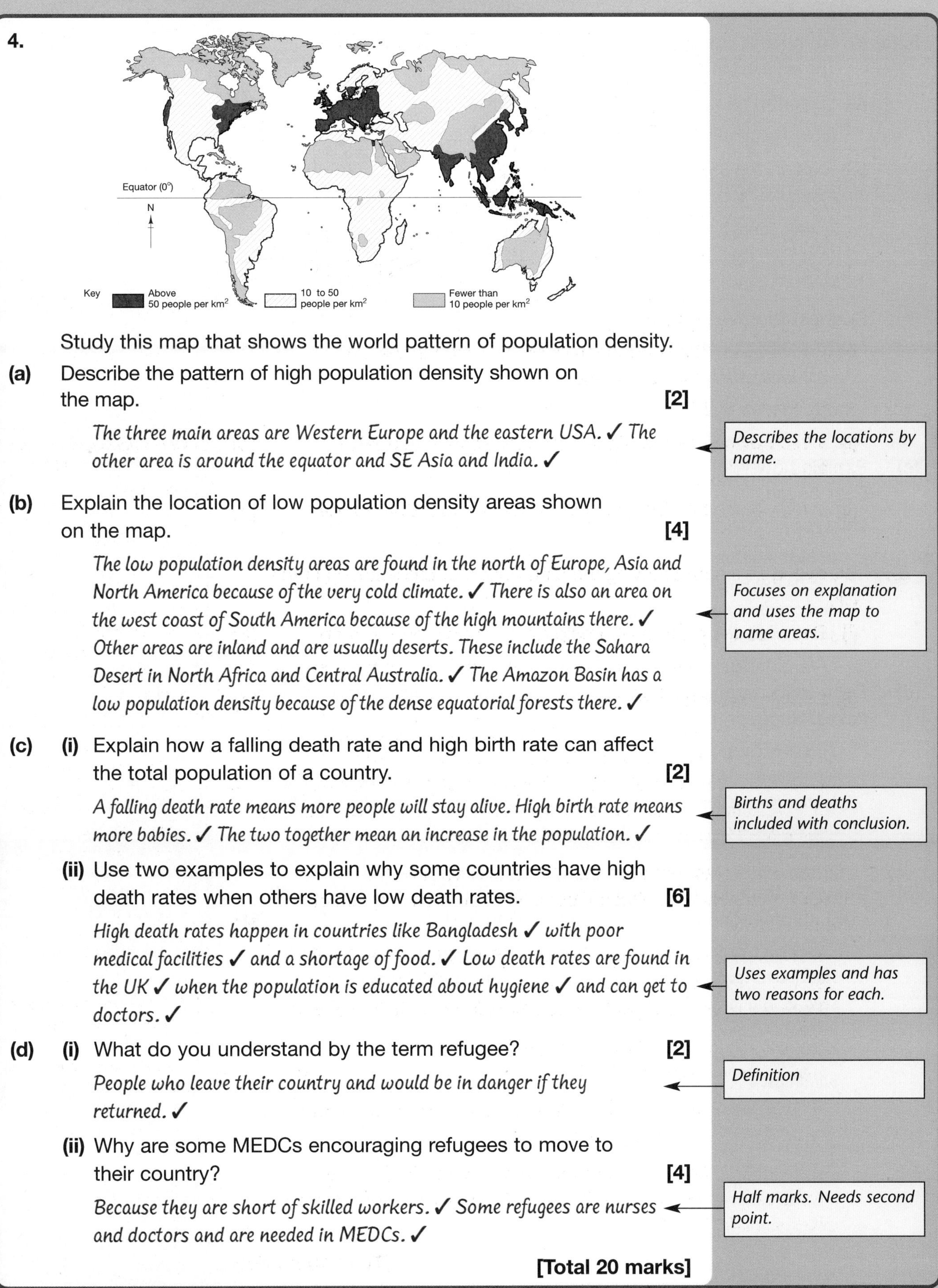

Study this map that shows the world pattern of population density.

(a) Describe the pattern of high population density shown on the map. **[2]**

The three main areas are Western Europe and the eastern USA. ✓ The other area is around the equator and SE Asia and India. ✓

> *Describes the locations by name.*

(b) Explain the location of low population density areas shown on the map. **[4]**

The low population density areas are found in the north of Europe, Asia and North America because of the very cold climate. ✓ There is also an area on the west coast of South America because of the high mountains there. ✓ Other areas are inland and are usually deserts. These include the Sahara Desert in North Africa and Central Australia. ✓ The Amazon Basin has a low population density because of the dense equatorial forests there. ✓

> *Focuses on explanation and uses the map to name areas.*

(c) **(i)** Explain how a falling death rate and high birth rate can affect the total population of a country. **[2]**

A falling death rate means more people will stay alive. High birth rate means more babies. ✓ The two together mean an increase in the population. ✓

> *Births and deaths included with conclusion.*

(ii) Use two examples to explain why some countries have high death rates when others have low death rates. **[6]**

High death rates happen in countries like Bangladesh ✓ with poor medical facilities ✓ and a shortage of food. ✓ Low death rates are found in the UK ✓ when the population is educated about hygiene ✓ and can get to doctors. ✓

> *Uses examples and has two reasons for each.*

(d) **(i)** What do you understand by the term refugee? **[2]**

People who leave their country and would be in danger if they returned. ✓

> *Definition*

(ii) Why are some MEDCs encouraging refugees to move to their country? **[4]**

Because they are short of skilled workers. ✓ Some refugees are nurses and doctors and are needed in MEDCs. ✓

> *Half marks. Needs second point.*

[Total 20 marks]

Sample GCSE question

5. (a) (i) Name one output from a dairy farm. [1]

The main output from a dairy farm is milk. ✓

No need for a sentence – just write 'milk'.

(ii) Describe two features of a market garden. [2]

Intensive farming ✓ *of high quality crops such as fruit and salad.* ✓

One mark for each.

(iii) Explain the terms 'milk quotas' and 'Common Agricultural Policy'. [4]

A milk quota is the maximum ✓ *amount of milk a dairy farmer is allowed to sell.* ✓ *The CAP is the way in which the EU plans its agriculture.* ✓

Second mark lost for CAP as explanation is a little vague.

(b) Suggest four reasons for the development of hill sheep farming in North Wales. [4]

In N Wales the mountains create slopes that are too steep, ✓ *the soils are thin* ✓ *and the climate too wet and cold* ✓ *for good crops.*

Four reasons at a mark each so be brief.

Climate point could be more specific.

(c) Explain how two of the following factors affect arable farming in East Anglia.

soil type   relief   climate   nearness to urban markets [4]

In East Anglia the land is fairly flat (relief) which means that machinery can be used easily. ✓ *Arable farming needs machinery such as ploughs and combine harvesters.* ✓ *London is a big market and is just to the south of East Anglia down the M11.* ✓✓

Good use of compass point to show direction and knows the number of the motorway.

(d) Farm trails, farm shops and caravan parks are being found on some farms. Suggest two reasons for this. [4]

This is because farmers have not been making enough money ✓ *and have had to find other sources of income.* ✓ *Also, people have more leisure time* ✓ *and look for ways to enjoy the countryside, like walking and caravans.* ✓

Need to be up-to-date with things that are happening around you.

(e) Describe the efforts that have been made to increase agricultural production in one LEDC that you have studied. [6]

The main effort to increase production from farms in LEDCs has been the Green Revolution. ✓ *This has used new high yielding varieties of seed* ✓ *and large amounts of fertiliser.* ✓ *Some governments have introduced more irrigation* ✓ *and paid agricultural advisers to talk to farmers.* ✓ *Governments have also given loans to buy machinery and buildings.* ✓

Description needed. Note the switch to LEDC.

A good well written answer. A pity no LEDC named where the Green Revolution has taken place.

[Total 25 marks]

Sample GCSE question

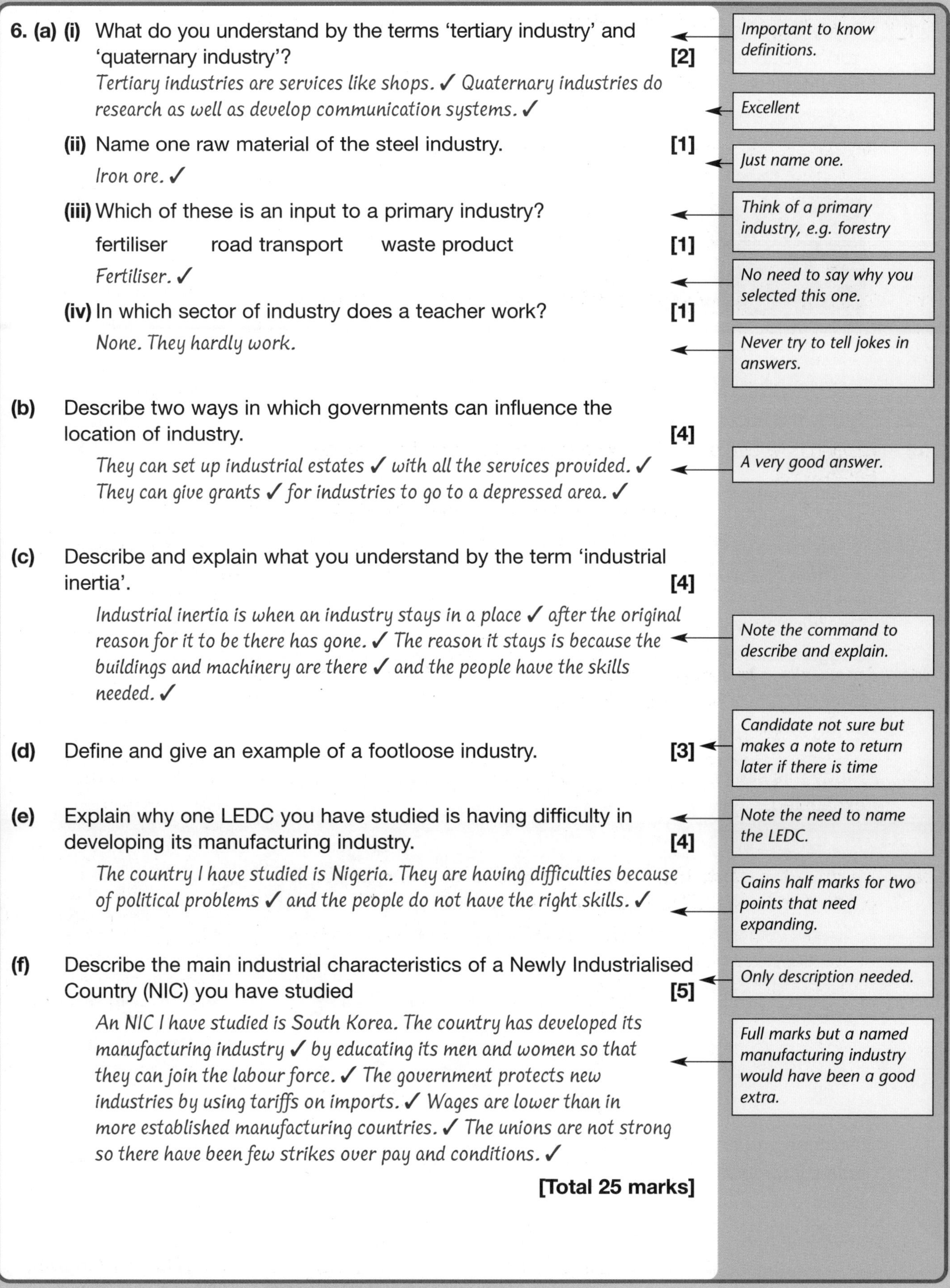

6. (a) (i) What do you understand by the terms 'tertiary industry' and 'quaternary industry'? **[2]**

Tertiary industries are services like shops. ✓ Quaternary industries do research as well as develop communication systems. ✓

> *Important to know definitions.*
>
> *Excellent*

(ii) Name one raw material of the steel industry. **[1]**

Iron ore. ✓

> *Just name one.*

(iii) Which of these is an input to a primary industry?

fertiliser　　road transport　　waste product **[1]**

Fertiliser. ✓

> *Think of a primary industry, e.g. forestry*
>
> *No need to say why you selected this one.*

(iv) In which sector of industry does a teacher work? **[1]**

None. They hardly work.

> *Never try to tell jokes in answers.*

(b) Describe two ways in which governments can influence the location of industry. **[4]**

They can set up industrial estates ✓ with all the services provided. ✓ They can give grants ✓ for industries to go to a depressed area. ✓

> *A very good answer.*

(c) Describe and explain what you understand by the term 'industrial inertia'. **[4]**

Industrial inertia is when an industry stays in a place ✓ after the original reason for it to be there has gone. ✓ The reason it stays is because the buildings and machinery are there ✓ and the people have the skills needed. ✓

> *Note the command to describe and explain.*

(d) Define and give an example of a footloose industry. **[3]**

> *Candidate not sure but makes a note to return later if there is time*

(e) Explain why one LEDC you have studied is having difficulty in developing its manufacturing industry. **[4]**

The country I have studied is Nigeria. They are having difficulties because of political problems ✓ and the people do not have the right skills. ✓

> *Note the need to name the LEDC.*
>
> *Gains half marks for two points that need expanding.*

(f) Describe the main industrial characteristics of a Newly Industrialised Country (NIC) you have studied **[5]**

An NIC I have studied is South Korea. The country has developed its manufacturing industry ✓ by educating its men and women so that they can join the labour force. ✓ The government protects new industries by using tariffs on imports. ✓ Wages are lower than in more established manufacturing countries. ✓ The unions are not strong so there have been few strikes over pay and conditions. ✓

> *Only description needed.*
>
> *Full marks but a named manufacturing industry would have been a good extra.*

[Total 25 marks]

Exam practice questions

1. Study this photograph of a 'tor' on Dartmoor, SW England.

(a) In the box, draw a sketch of a tor. Add detailed labels to show the characteristics of a tor and describe how it was formed. **[4]**

(b) Explain how tors may have been formed. **[3]**

....

....

2. Study the Fact File on Whatley Quarry, in the Mendip Hills Somerset.

Fact File

- Limestone is quarried here
- At current working levels the quarry will close in seven years
- There is a workforce of 100 people
- The quarry has a rail link to transport crushed rock to customers
- Whatley is one of the most modern and efficient quarries in Europe

The quarry owners have applied for planning permission to extend the quarry. Suggest and explain two reasons why planning permission might be given. **[4]**

....

....

3. Study this sketch map showing proposals to restore the quarry site when reserves are exhausted.

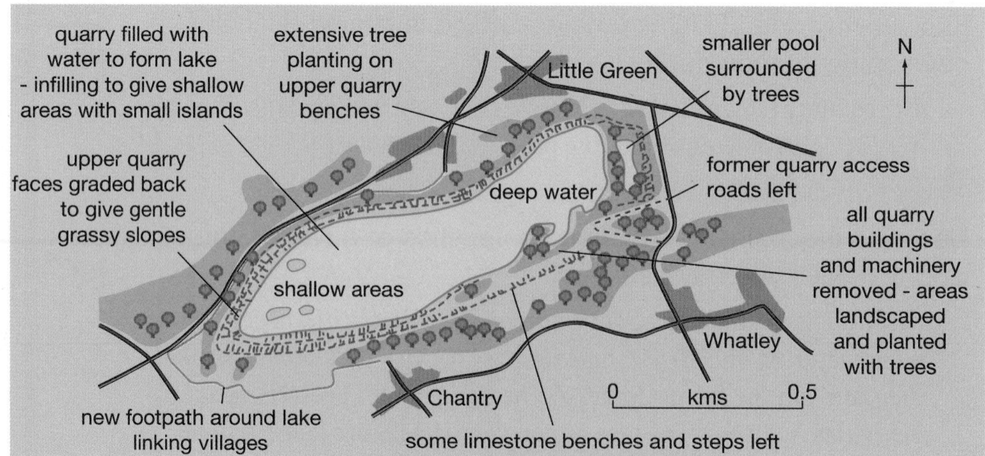

Describe and explain **two** ways in which the restored site might be developed as an amenity for local people. **[4]**

....

....

[Total 15 marks]

Exam practice questions

1. This diagram shows a cross-section of a river valley in a lowland area.

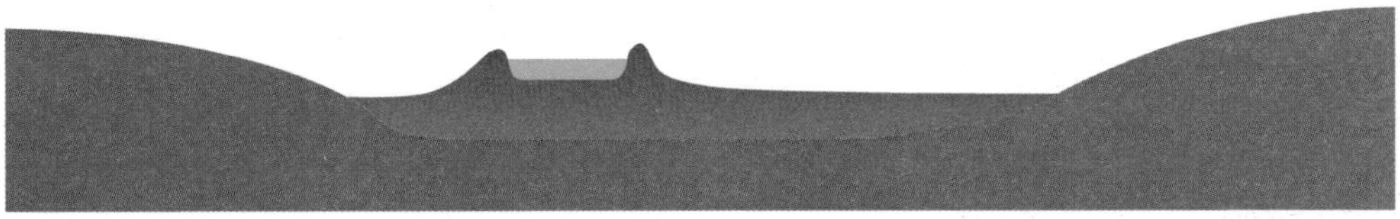

(a) On the diagram, clearly label:
the floodplain levées river **[3]**

(b) Explain how man-made levées can prevent flooding. **[2]**

....

....

2. With the aid of a sketch map and detailed labels

(a) name a river delta you have studied

(b) explain the formation of a delta. **[6]**

3. Describe and explain the **human** causes of a flood you have studied in an LEDC. **[4]**

....

....

....

[Total 15 marks]

Exam practice questions

1. **(a)** Which two of the following are most likely to be found in the Central Business District?

large department store — main bus station

secondary school — football stadium **[2]**

(b) For a CBD you have studied, describe and explain two changes taking place in the area. **[4]**

....

....

2. Study the diagram of the concentric zone model.

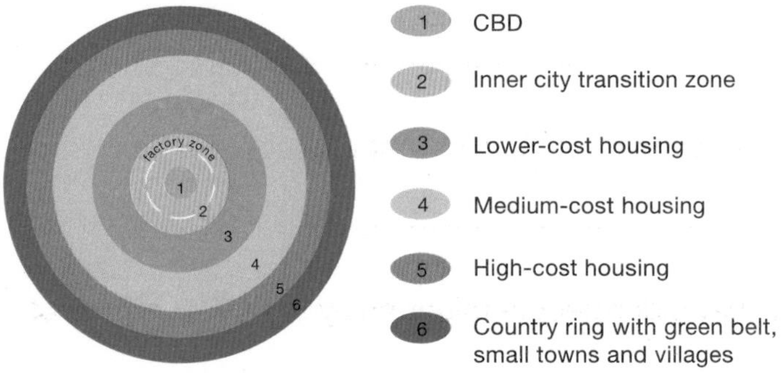

factory zone

1 CBD

2 Inner city transition zone

3 Lower-cost housing

4 Medium-cost housing

5 High-cost housing

6 Country ring with green belt, small towns and villages

(a) Compare the housing in zones 3 and 5. **[4]**

....

....

(b) Modern factories are moving to zone 6. What are the advantages of this new location? **[3]**

....

(c) Describe two advantages and two disadvantages of having a green belt around an urban area. **[8]**

....

....

....

3. Give two reasons for the growth of shanty towns around cities in LEDCs. **[4]**

....

....

[Total 25 marks]

Exam practice questions

1. **(a)** The use of coal for energy production in the UK has fallen in recent years. Suggest two reasons for this. **[2]**

....

(b) Suggest two reasons for the increased use of natural gas for energy production in the UK in recent years. **[2]**

....

(c) Describe two advantages and two disadvantages of developing sources of hydroelectric power. **[4]**

....

....

2. **(a)** Identify two physical features of the Lake District that attract tourists. **[2]**

....

(b) Name two types of holiday that older people might enjoy in the Lake District. **[2]**

....

(c) Describe two advantages and two disadvantages that tourists bring for people living in the Lake District. **[4]**

....

....

3. Global warming could be worse in 100 years.

(a) Describe how global warming is said to be taking place. **[2]**

....

(b) Describe two ways in which individuals could help reduce the cause of global warming. **[2]**

....

4. Some governments have increased their national energy production by using nuclear power. Describe the evidence against this strategy. **[5]**

....

....

[Total 25 marks]

Exam practice questions

Study the table below of data about two countries.

Measures of development			
Country	**Life expectancy (years)**	**Adult literacy rate (%)**	**People per doctor**
Ethiopia	47	33	38000
Japan	79	99	613

1. **(a)** Explain how the information in the table suggests that Japan is an MEDC. **[4]**

   ....

   ....

   (b) Suggest two other measures of development not shown in the table. **[2]**

   ....

2. Describe the different problems of water supply in LEDCs and MEDCs. **[6]**

   ....

   ....

   ....

3. **(a)** Describe the difference between imports and exports. **[2]**

   ....

   (b) Explain two reasons for some LEDCs having difficulty in increasing their exports. **[4]**

   ....

   ....

4. **(a)** Describe the different purposes of short term and long term aid. **[4]**

   ....

   ....

   (b) Describe one way in which aid can help an LEDC achieve sustainable development. **[3]**

   ....

[Total 25 marks]

Additional skills questions

1. Look at the figure below showing percentage use of rock aggregates in the UK in 2001.

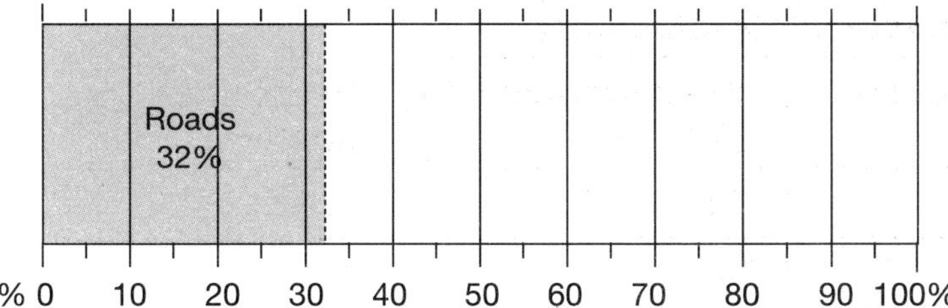

Use	%
Roads	32
Housing	25
Offices and shops	14
Factories and warehouses	13
Other public works	16

(a) Complete the graph using the data in the table. **[4]**

(b) One kilometre of new motorway requires about 125 000 tonnes of aggregate. The M6 Toll Road is about 32km long. Approximately how many tonnes of aggregate were required? **[1]**

....

2. The graph below shows climate figures for Archangel (latitude: $64\frac{1}{2}$°N).

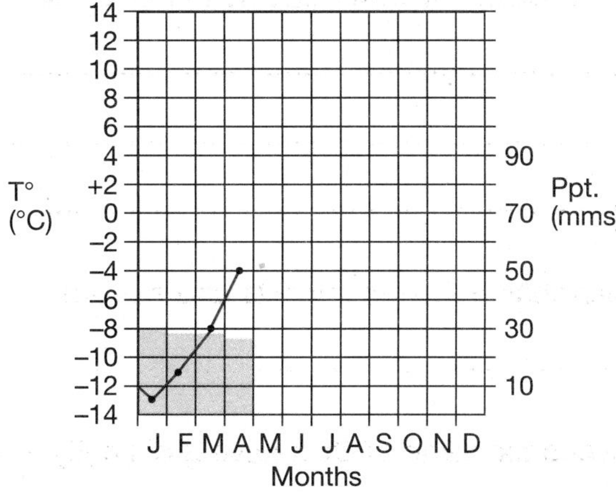

Month	J	F	M	A	M	J	J	A	S	O	N	D
T (°C)	–13	–11	–8	–4	4	10	12	10	3	–4	–8	–10
Ppt. (mm)	30	27	27	24	40	50	69	76	60	50	45	40

(a) Complete the graph using the data in the table. **[4]**

(b) **(i)** What is the temperature range?

....

(ii) What sort of precipitation most likely falls from October to April?

....

(iii) What type of trees have adapted to this severe climate?

.... **[3]**

[Total 12 marks]

Additional skills questions

1. Study the two photographs of a honeypot site in the Lake District National Park.

(a) **(i)** Explain the meaning of the term 'honeypot site'. **[2]**

....

(ii) Describe two pieces of evidence from each photograph to suggest that the honeypot site is successful. **[4]**

....

(b) Describe two problems that face local people living in this honeypot site. **[2]**

....

(c) Describe two ways in which the problems you have described can be reduced. **[2]**

....

2. Study the two population pyramids shown.

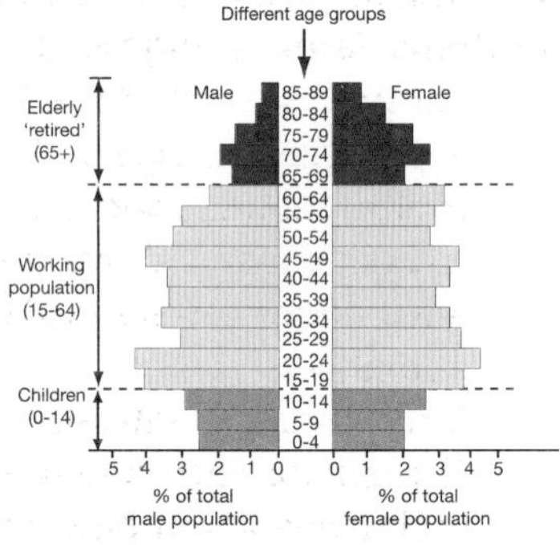

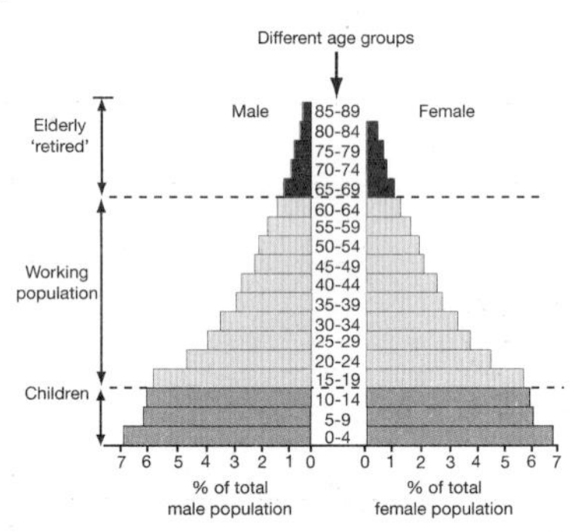

(a) Which pyramid is of the population of an LEDC? .... **[1]**

(b) In which pyramid is life expectancy highest? .... **[1]**

(c) On a population pyramid, which groups form the dependent population? **[2]**

....

(d) Describe two reasons for death rates to be falling in most countries. **[2]**

....

(e) The populations of MEDCs are often ageing. Describe two problems that may result from this. **[4]**

....

....

[Total 20 marks]

Answers

Ordnance Survey map questions

1. (a) Cairn
   (b) (i) North (ii) 2km (iii) Ridge (steep sided)
   (iv) Cliffs
   (Remember – north is always at the top of the map. Use the scale to work out the distance. Some landforms such as (iv) are shown in the key. For others you need to 'read' the contours. Some types of land use are also shown in the key.)
2. (a) Mixed woodland
   (b) 2.1km (allow 2.0 to 2.3)
   (c) Short level walk through the wood, then steeply up following the stream. Leave woods for open moorland – steep climb up to 400m contour. Less steep for just over 0.5km, before slope steepens again with a short, very sharp rise to the summit. Quite a demanding walk!
   (1 mark for each change. Reserve 2 marks for change in land use.)
3. (a)

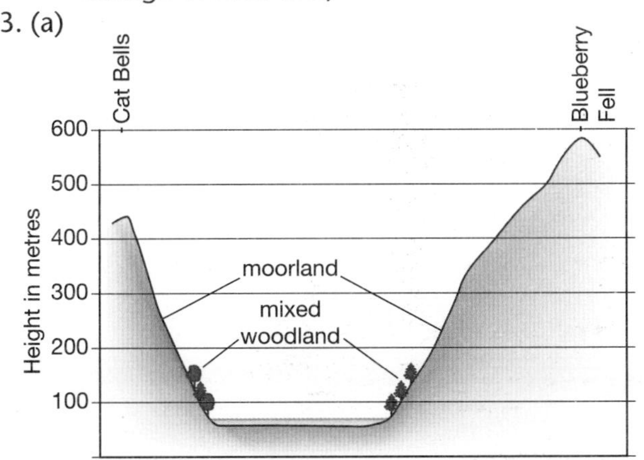

   (2 marks for accuracy, 1 mark for labelling 'Blueberry Fell')
   (b) About 1:2 (a very steep slope) (750/381)
   (Gradient is measured by dividing horizontal distance/difference in height, e.g. 600/200 = 1:3)
   (c) Wide, flat valley floor – very steep valley sides rise sharply from the floor.
   (Any three points. Credit detailed map evidence, e.g. valley floor is 5km wide.)
   (d) Woodland on both slopes from the lakeside up to about 150–175m – moorland above both woodland areas.
   (1 mark each for 3 clearly labelled land uses –see completed cross section.)
4. The Derwent has a broad flat-floored valley compared to the steep-sided V-shaped valley of Whit Beck. The Derwent has a winding course with some evidence of oxbow lakes (GR 248247), but Whit Beck is much straighter and falls more steeply. The River Derwent is a larger river than Whit Beck.
   (1 mark for each comparative point)
5. (a) Caravan park, camping ground, hotel, cycle network, marked path (any 2, worth 1 mark each)
   (b) 264236 (allow 1 either way in 3rd and 6th figures)
   (c) National Trust Stone Circle (allow cycle route)
6. (a) One major E to W route, one N to S route, small number of minor roads, focus on Keswick. (3 points worth 1 mark each)
   (b) Railway now dismantled but signs of embankments and cuttings and circuitous route (1 located piece of evidence = 2 marks)
7. South
8. One major town, settlement mainly in the valley bottoms, farms in highland valleys.
   (2 clear points worth 2 marks each)
9. (a) Bypass built, congested road pattern at centre, roundabouts and T-junctions
   (2 points worth 2 marks each)
   (b) Advantages: leisure facility, bed and breakfast potential
   Disadvantages: intrusion, noise, litter, crime
   (1 mark for each description and 1 mark for each explanation plus 2 marks for map evidence)

Exam practice questions

Rocks and landscapes

1. (a) Your sketch needs to fill the box and clearly show the shape of the rocks and the deep joints that give the tor its characteristic form. Labels should include: hilltop location; rounded granite rocks, often becoming smaller towards the top; deeply weathered joints.
   (1 mark for adequate sketch, 1 mark for each label = 3)
   (b) *Formation*:
   'Tor' formed under the ground; deep chemical weathering during warmer climates in the past caused granite to disintegrate

Joints widened; solid blocks of granite left, surrounded by loose weathered rock
At the summits of hills the weathered rock has been eroded away to leave an outcrop of rounded granite rocks
NB Scientists still disagree as to how tors are formed.
(Three developed points. Credit understanding of the stages [sequence] of formation.)

2. Use the Fact File, but develop *both* reasons:
Limestone aggregate is very important in the construction industry and many others (reason); need to maintain a reliable supply (develop).
Maintain employment in a rural area (reason) for more than seven years (develop).
The rail link reduces road transport (reason); less pollution and dust in local villages (develop).
Keeping modern, efficient quarries keeps prices low (reason) and will allow the closure of many small, less efficient quarries (develop).
(1 mark for each reason, max 2. 2 marks for each development.)
3. Use the resource to develop ideas such as:
Water sports (1 mark) – fishing, sailing, wind-surfing are popular sports (1 mark)
Nature reserve (1 mark) – varied habitats, bird hides, interpretation centre (1 mark)
Picnic places/barbeque sites, café
Bicycle hire, jogging/fitness circuit
(1 mark for each 'way' described, max 2. 1 mark for explanation.)

River landscapes

1. (a) Your labels must clearly identify the three features asked for (see page 29, fig 2.12).
(b) Levées allow more water to be held in the river channel.
Levées can be raised to contain previous known flood levels.
2. Use an example of a delta you have studied.
Clearly draw the delta and name the river.
Labels to explain the *formation* should include:
Large river carrying heavy loads of silt
River enters the sea, loses energy and deposits large amounts of silt
River splits up around the banks of deposition to form distributaries
Sheltered coast, sea unable to remove all the silt
Distributaries spread the silt out into the sea to form a delta
(2 marks for naming a river delta and providing a reasonable sketch map. 1 mark [max 4] for each explanation point – credit understanding of the sequence of events.)
3. Make sure you select an example from an LEDC.
Link causes *clearly* to the example you have chosen (use place names).
Name the LEDC and flood event (river? date?)
Human causes of a flood in a LEDC may include:
Deforestation for agriculture (population growth)
Illegal logging (tropical rain forest)
Overgrazing and soil erosion of levees (population growth)
Flood prevention schemes upstream (problems of multi-national drainage basins)
Lack of investment in flood prevention schemes (low GDP – alternative priorities)
(2 marks for each developed cause. Max 2 marks if LEDC not named. No credit for physical causes.)

Settlement

1. (a) large department store
main bus station (1 mark each)
(b) Description: Rebuilding brownfield sites with housing, road building schemes, renewal of commercial buildings. (2 changes described)
Reasons: Changes reflect pressure on the area to stay attractive, alive and up-to-date. (2 reasons given)
2. (a) Zone 3 has terraced houses with small gardens or yards.
Zone 5 has detached houses with gardens and car parking.
(2 marks for each comparison)
(b) New premises (1), more space (1), better access (1), closer to workers (1)
(c) Advantages include: prevents sprawl, green areas near homes, space for recreation.
Disadvantages include: increases commuting time, increases house prices, constrains good planning schemes.
(Advantages: 2 at 2 marks = 4;
Disadvantages: 2 at 2 marks = 4)
3. Too many people moving into the area, lack of housing, poor people. (2 marks for each developed reason)

Managing resources

1. (a) Causes pollution; reserves depleted; cleaner alternatives; closure of mines; high costs; trade unions (2 reasons worth 1 mark each)
(b) Resource has become available; easier to transport; cleaner than coal (2 reasons worth 1 mark each)
(c) Advantages: sustainable, clean, relatively cheap, water areas for recreation
Disadvantages: costly to build, danger of flooding, loss of river water, attracts malaria, may silt up
(1 mark for each advantage, 1 mark for each disadvantage)
2. (a) Dramatic scenery; beautiful lakes; glaciated valleys; upland tarns (2 features worth 1 mark each)
(b) Painting; walking; guided tours; lake and rail trips; heritage; town shops (2 types worth 1 mark each)

(c) Advantages: income, work, keeps shops open, tourist facilities, businesses
Disadvantages: congestion, higher house prices, noise, fewer homes for locals
(1 mark for each advantage, 1 mark for each disadvantage)

3. (a) Production of greenhouse gases, trapped in the atmosphere, reflect heat back to the earth, raise temperatures (2 reasons at 1 mark each)
(b) Increase recycling, fewer car journeys, holidays taken in the UK, use of recyclable materials and better house insulation (2 ways at 1 mark each)

4. Disasters such as Chernobyl and Five Mile Island.
Problems of disposal of waste.
Threat of terrorist activity.
Sustainable alternatives being developed.
Public concern about safety.
(1 mark for each well-developed point)

Development

1. (a) Higher life expectancy shows better quality of life, education available, people have access to doctors (1 mark for each indicator plus 1 for clear link to MEDC characteristics)
(b) Birth rates, death rates, infant mortality, car ownership, industrial structure (2 indicators worth 1 mark each)

2. LEDCs: lack of clean water, irregular supply, areas not served
MEDCs: increased costs, availability in some areas, water shortages affect certain businesses, e.g. garden centres, car washes
(3 marks for LEDCs and 3 for MEDCs)

3. (a) Imports: goods purchased abroad and brought into a country
Exports: goods purchased by other countries and sent to them. (1 mark for each)
(b) Poor management, lack of capital, corruption, poor transport services, trade barriers, failed harvests, high local demand (2 reasons at 2 marks each)

4. (a) Short term: emergency, follows natural disaster, to meet immediate need
Long term: education and training to produce lasting improvements, usually a project
(2 marks for each)
(b) Use of appropriate technology, aimed at local people, not dependent on experts from developed countries to sustain project, low tech, low skill, not bureaucratic, local people in control (1 way linked to sustainability worth 3 marks)

Additional skills questions

Physical geography

1. (a)

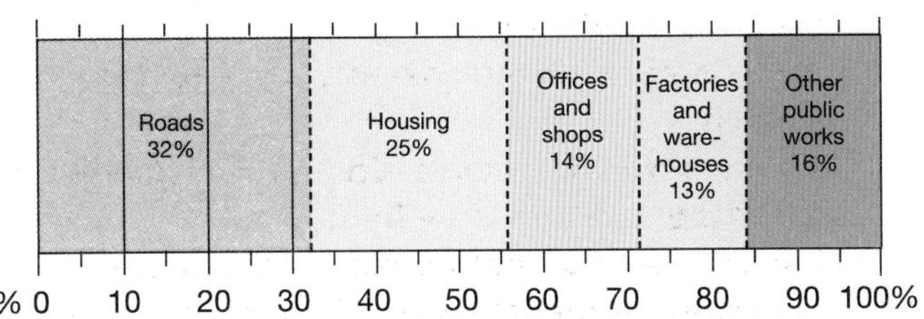

(3 marks for accuracy of graph and 1 mark for adding labels correctly)
(b) 4m tonnes

2. (a) Copy the graph onto your graph paper (see page 85, Fig 6.7A)
(2 marks for accuracy in each part – temperature and precipitation)
(b) (i) 25 °C (ii) Snow (iii) Coniferous

Human geography

1. (a) (i) Popular tourist site, very crowded in summer, facilities managed to meet large demand.
(ii) Crowds of people, facilities in use, coach parties, hotel in background, congested shopping street, people in holiday clothes
(2 descriptions worth 2 marks each)
(b) Problems include: difficult parking, increased noise, increased house prices, shops' focus on tourists, unemployment off-season (1 mark for each problem)
(c) Parking restrictions and permits schemes for locals, affordable housing for locals only, pedestrian only areas, extension of season with other attractions (2 ways outlined at 1 mark each)

2. (a) The right hand pyramid
(b) The left hand pyramid
(c) Young people before working (1), the older retired people (1), others not working not shown
(d) Better health care, better diet, cleaner living conditions, improved water supply, access to doctors, more widespread education
(2 reasons worth 1 mark each)
(e) Increased burden on the working population, higher welfare costs, higher state pension costs, more hospitals and doctors needed, larger houses occupied by older couples
(2 problems worth 2 marks each)

Index